Living in the Plastic Age

Johanna Kramm and *Carolin Völker* are research scientists at ISOE – Institute for Social-Ecological Research in Frankfurt. From 2016 to 2021, they led the interdisciplinary SÖF junior research group PlastX, researching plastics in the environment from a social-ecological perspective.

Johanna Kramm, Carolin Völker (eds.)

Living in the Plastic Age

Perspectives from Humanities, Social Sciences
and Environmental Sciences

Campus Verlag
Frankfurt/New York

The work, including all its parts, is subject to copyright protection. The text of this publication is published under the "Creative Commons Attribution-ShareAlike 4.0 International" (CC BY-SA 4.0) licence.
The full licence text can be found at:
https://creativecommons.org/licenses/by-sa/4.0/deed.en

Any use that exceeds the scope of the CC BY-SA 4.0 licence is not permitted without the publisher's consent.
The images and other third-party material contained in this work are also subject to the aforementioned Creative Commons licence, unless otherwise stated under references/illustration index. Insofar as the material in question is not subject to the aforementioned Creative Commons licence and the usage in question is not permitted under statutory provisions, the consent of the respective rightsholder must be obtained.

ISBN 978-3-593-51445-1 Print
ISBN 978-3-593-44902-9 E-Book (PDF)
DOI 10.12907/978-3-593-44902-9

Copyright © 2023 Campus Verlag GmbH, Frankfurt am Main
Some rights reserved.
Cover design: Campus Verlag GmbH, Frankfurt am Main
Cover illustration: Vesna Nestorovic
Typesetting: Garamond
Printing office and bookbinder: Beltz Grafische Betriebe GmbH, Bad Langensalza
Beltz Grafische Betriebe GmbH ist ein klimaneutrales Unternehmen (ID 15985-2104-1001).
Printed in Germany

www.press.uchicago.edu
www.campus.de

Contents

Introduction: Living in the Plastic Age...7
Johanna Kramm, Carolin Völker

Explaining Agenda-Setting of the European Plastics Strategy.
A Multiple Streams Analysis..25
Paula Florides, Johanna Kramm

Microplastics in the Aquatic Environment..51
Maren Heß, Carolin Völker, Nicole Brennholt, Pia Maria Herrling, Henner Hollert, Natascha Ivleva, Jutta Kerpen, Christian Laforsch, Martin Löder, Sabrina Schiwy, Markus Schmitz, Stephan Wagner, Thorsten Hüffer

Risk Perception: The Case of Microplastics. A Discussion of
Environmental Risk Perception Focused on the Microplastic Issue...........87
Marcos Felipe-Rodriguez, Gisela Böhm, Rouven Doran

Everyday Life with Plastics: How to Put Environmental Concern into
Practice(s)..111
Immanuel Stieß, Luca Raschewski, Georg Sunderer

Using Citizen Science to Understand Plastic Pollution: Implications
for Science and Participants..133
Marine Isabel Severin, Alexander Hooyberg, Gert Everaert, Ana Isabel Catarino

Behavior Change as Part of the Solution for Plastic Pollution....................169
Maja Grünzner, Sabine Pahl

Chemicals in Plastics: Risk Assessment, Human Health Consequences, and Regulation 197
Jane Muncke, Lisa Zimmermann

Circularity in the Plastic Age: Policymaking and Industry Action in the European Union 235
Sandra Eckert

Plastivores and the Persistence of Synthetic Futures 261
Kim De Wolff

Contributors 277

Index 283

Introduction: Living in the Plastic Age

Johanna Kramm, Carolin Völker

Living in the Age of Plastics

The notion Plastic Age draws on epochal categories such as Stone Age and Iron Age. These notions describe periods of time in which the specific material shape the social artefacts and technologies in such profound ways that the materials become characteristic for the societies in that periods (Bensaude Vincent 2013).

The notion of Plastic Age therefore implies that a material and its materiality with all its relations it engages enables a new way of living. Materials matter, but they need to be sensed in a relational way. Plastics materialize through practices and technological configurations, are symbolically loaded, and enable new economic, technological, natural, and societal relations all at once.

The first plastic materials were invented in the 19th century to substitute other materials like wood, horn or metals. At this time first plastics precursors like Parkesine created by Alexander Parkes around 1860 or celluloid developed by John Wesley and Isaiah Hyatt in the 1870s were praised as solutions to dooming scarcities of resources such as tortoise shell and ivory and considered as means for relieving the pressure on natural resources (Meikle 1995). About 100 years later, plastics have evolved from a "symbol of progress and modernity" (Umüßig 2021) into a symbol of mass consumption, a society of disposables, and unmanageable amounts of waste with far-reaching consequences for the environment. The accumulation of plastic waste in the oceans illustrates these consequences and is considered as one of the most urgent environmental problems today.

While Robert Sklar (1970) identifies the start of the Plastic Age after the First World War with the first developments of mass culture, the cultural historian Jeffrey Meikle (1995) argues that plastics became to play a crucial role in American culture in the second half of the twentieth century. As Bensaude Vincent (2013, 19) has argued, these temporal positionings show that synthetic materials did not revolutionize economies, cultures, and society all at the same time. However, about 150 years later, we can say, synthetic polymers have found their way into all realms of life, from daily routines of hygiene, cooking or shopping, to transport, construction, leisure, agriculture or medical care (PlasticsEurope 2020). A reason for their versatile applications is that "plastics" refer not to a single material, but to a range of different polymers with different additives resulting in materials with different properties and capabilities. In the process of polymerization, the plastic product is molded and can take any shape. Additives help to create multiple properties. For instance, plasticizers can make plastics flexible, colorants can give the product any desired color, thermal stabilizers create heat resistance, and flame retardants enhance flame resistance.

The problems related to plastics are closely intertwined with the establishment of mass consumption after the Second World War. Single-use products, such as all kinds of packaging, created an unprecedented waste problem. Additionally, plastics have been found to be problematic due to their building components like bisphenol A or additives like phthalates. It has been proven that these chemicals interfere with the hormone system, and so some of them are now banned in certain products. During the last decades, attention was also drawn to finely fragmented plastic particles, commonly known as "microplastics", that are traceable even in remote locations such as the Arctic (Bergmann et al. 2022).

Information about the downsides of plastics has exploded in recent years and has been covered in various books, documentaries, museum exhibitions, as well as by environmental non-governmental organizations, coalitions, and global campaigns on microplastics, single-use plastics, and zero-waste (e.g., "Beat the Microbead" or "#BeatPlasticPollution"). As a result, an increasing awareness has arisen about the implications of plastic use, such as microplastics (e.g., for risk perception in Germany see Kramm et al. 2022) with a consequential public debate about measures to curb plastics usage. A further result has been the development of policies and strategies such as the European Strategy on Plastics. Moreover, several funding lines on plastics were launched and possible measures for a sustainable use have emerged (e.g., for Europe JPI Oceans and H2020 on marine plastics and litter and another one on microplastics and human health risks, for Germany the funding line "Plastics in the Environment" by the Federal Ministry for Education and Research). Global attention to plastics as a pressing environmental issue has led the international community to endorse a resolution at the United Nations Environment Assembly in May 2022 to establish a legally binding agreement to "end plastic pollution" by the end of 2024 (UNEA 2022).

This resolution demonstrates that it is now globally recognized that the plastic economy needs to be fundamentally transformed in order to enable a more sustainable use of plastic materials. This transformation should not only include measures to prevent plastic waste from entering the environment, but must also cover other aspects such as production and consumption. Nevertheless, plastics remain the "workhorse" of the global economy (WEF 2016). Since 2016 global plastics production increased from 322 to almost 370 tons per year (PlasticsEurope 2020) with a market size of approximately 580 billion U.S. dollars in 2020 (Statista 2021). A sustainable future use of plastics therefore requires a transformation of the linear economy towards more circularity and sufficiency. This change is a major challenge and requires the cooperation of science and societal actors at all levels. However, the idea of transformation should not solely focus on or stop at the economy, but should permeate all societal realms. This thought informs the rationale for this anthology. Since plastics permeates not only all societal realms but also all ecosystem and natural realms as well, we need to scrutinize plastics from multiple angles. This anthology, therefore, compiles research from different disciplines ranging from humanities, social sciences, psychology, political sciences and environmental sciences.

Background of the Book: Inter- and Multidisciplinary Perspectives

As the issue of plastics is part of both scientific and public debate, science is called upon to contribute to questions raised in public discourse with empirical and normative research. Problems associated with the production, use, and disposal of plastics are complex, as is the case with most sustainability issues, and require knowledge from multiple disciplines as stated above. For instance, plastics in the environment stem from numerous sources and are hardly traceable back to their origin. Due to the diversity of environmental plastics and the resulting biological effects, it is difficult to quantify the risks to the environment and human health. In addition, plastics are so closely intertwined with almost all aspects of everyday life that the topic encompasses an inexhaustible wealth of societal actors, practices, technological and policy fields, each bringing different perspectives and interests to the table.

Our research group "PlastX—Plastics in the Environment as a Systemic Risk"[1], which we led from 2016 until 2021, took an inter- and transdisciplinary approach and looked at the problem in a holistic way. The PlastX team members shared the common boundary object of "plastics in the environment" and examined the problem from different disciplinary angles. Ecotoxicologists analyzed the environmental risks of microplastics (Zimmermann et al. 2020b), while human geographers focused on the science-policy interface (Kramm 2022). Furthermore, human geographers studied the governance mechanisms of marine plastic pollution (Kerber and Kramm 2021; 2022). In interdisciplinary collaboration, the translation of scientific findings on the risks of microplastics into public discourse as well as public risk perception was investigated (Kramm et al. 2022; Völker et al. 2020). To elucidate solution strategies that cover the supply chain of a plastic product, another focus was placed on plastic packaging. Sociologists analyzed the role of plastics in food supply practices (Sattlegger et al. 2020; 2021a; 2021b), chemists evaluated bioplastics as an alternative to conventional plastics (Haider et al. 2019), and ecotoxicologists investigated the toxicity of chemicals in both conventional and bioplastics (Zimmermann et al. 2019; 2020a; 2021).

The holistic perspective highlighted the multiple problems associated with plastics and subsequently the complexity of the issue. By framing the problem as a "systemic risk" (Klinke and Renn 2006), its complexity was addressed as well as the following further characteristics, such as ambiguity or divergent perceptions, high uncertainty of scientific data, and the

---

1 PlastX was funded by the German Federal Ministry for Education and Research (BMBF) as part of the funding priority "SÖF – Social-ecological research" within the funding area "Junior research groups in social-ecological research". The group included the (former) PhD students Dr. Lisa Zimmermann (ecotoxicology), Dr. Tobias Haider (chemistry), Dr. Lukas Sattlegger (sociology) and Heide Kerber (human geography) as well as the PostDocs and project coordinators Dr. Johanna Kramm (human geography) and Dr. Carolin Völker (ecotoxicology).

everyday production of environmental risk of plastics (Kramm and Völker 2018). Moreover, the inter- and transdisciplinary collaboration included the perspectives and priorities of various societal actors as for instance industry, retailers, water authorities, nature conversation, development cooperations, and consumer protection groups. Thus, the PlastX team members were able to gain a joint and more detailed understanding of the issue and to critically evaluate solution strategies and their potential implications (Kramm et al. 2020).

The members of the PlastX team also held lectures and seminars at Goethe University Frankfurt where they shared the results of the interdisciplinary collaboration

The broad network was used to organize the public lecture series "Living in the Plastic Age" in summer semester 2019 with speakers from the humanities, social sciences, and natural sciences. With their insights into different disciplinary approaches, the various experts met with broad interest from the auditorium that also included students from various disciplines. The book "Living in the Plastic Age" ties in with the lecture series and aims to spark interest among a broad readership, highlight different approaches and perspectives on plastics, and encourage interdisciplinary collaboration. The anthology presents the multidimensional facets of plastics and microplastics from different disciplinary angles including political and environmental sciences, psychology, sociology, ecotoxicology, environmental humanities, and science and technology studies (Tab. 1). While some chapters are taking a disciplinary perspective, others are interdisciplinary, i.e., integrating different disciplines.

Tab. 1: *Chapters of the anthology "Living in the Plastic Age" and the respective disciplines.*

Chapter	Disciplin(es)
(1) Explaining Agenda-Setting of the European Plastics Strategy. A Multiple Streams Analysis	Political Sciences
(2) Microplastics in the Aquatic Environment	Environmental Sciences
(3) Risk Perception: The Case of Microplastics. A Discussion of Environmental Risk Perception Focused on the Microplastic Issue	Psychology
(4) Everyday Life with Plastics: How to Put Environmental Concern into Practice(s)	Sociology
(5) Using Citizen Science to Understand Plastic Pollution: Implications for Science and Participants	Environmental Sciences, Environmental Psychology, Sociology
(6) Behavior Change as Part of the Solution for Plastic Pollution	Psychology
(7) Chemicals in Plastic Packaging: Challenges for Regulation and the Circular Economy	Ecotoxicology
(8) Circularity in the Plastic Age: Policymaking and industry action in the European Union	Political Sciences
(9) Plastivores and the Persistence of Synthetic Futures	Environmental Humanities, Science and Technology Studies

Overview of the Book

As is evident from Table 1, the chapters of the book are not arranged by discipline, but in thematic order. The anthology opens with different descriptions of the problem, from both political and natural science perspectives, and then addresses social aspects, risk perception, and behavior change. It then moves on to harmful chemicals, circular economy, and a critical perspective on petro-capitalism. In the following, an overview of the chapters is given.

In the last decades plastics have achieved a "new" political status. While previously under waste legislation, plastics were treated as one waste material amongst others, they now have a status of their own with special strategies and policies.

In the contribution on agenda-setting of the European plastics strategy, *Paula Florides and Johanna Kramm* trace how plastics were scientifically and politically problematized. They show how the topic of plastics was subsequently shifted from a long-known environmental problem that was addressed in European legislation only in a scattered way, to a prominent place on the agenda of the European commission, a development that has enabled the material to be addressed in a more systematic way. They reconstruct the way it took to put plastics onto the political agenda. Using the Multiple Streams Framework, the authors analyze how different streams such as problem frames, policies and policy ideas and political institutions, e.g., governments, the European Commission or pressure-groups, are coupled during windows of opportunity. Agenda-setting is then a result of the coupling of these different streams (problem, policy and politics). Environmental concern about plastic waste, especially the one in oceans, has been raised at least since the 1970s (Kramm and Völker 2018). But it was not until the early 2000s with the observation of plastic debris accumulations in the pacific gyres by Charles Moore (Moore et al. 2001) and the coining of the term "microplastics" by Thompson et al. (2004) that the scientific and public debate on plastics in the environment accelerated. As research on policy-processes has pointed out (e.g., Kingdon 2014), "problems" have to be framed in such a way that policy-makers recognize the need to respond to it. In the case of plastics, pressure groups, some of which were closely collaborating with scientists (e.g., Algalita, 5 Gyres, Beat the Microbead) and the media contributed to a problem framing that put a certain pressure on policy-makers. The authors show that marine litter was not explicitly connected to plastics but was rather considered as something to be dealt with by the coastal member states of the EU or in the context of international agreements on marine pollution. A "Europanization" of the problem raised political concerns and created a sense of responsibility among European policy-makers. This was achieved by two shifts: the first concerned the discourse on marine litter that changed from a sea-based perspective to the inclusion of land-based sources connecting marine plastics to waste management policies and the second was a shift from spatial distance to spatial proximity by detecting plastic debris and microplastics along European coasts and in European rivers and lakes. Moreover, the former European Commissioner of the Environment played an important role in putting plastics onto EU's agenda by connecting plastic waste to his ongoing plans of resource efficiency that paved the way for the European plastics strategy. The analysis shows that environmental problems need to be made explicit and that proposals for policy are equally necessary.

The plastics strategy was adopted in 2018 and is a key element in the European circular economy action plan. Next to measures to curb plastic waste, improve plastics recycling, and foster innovation and international development, measures to reduce microplastics are another central part of the strategy. Meanwhile, microplastics are found ubiquitously in the environment and in the last decades a new field of science has developed with its own research methods, journals (Microplastics and Nanoplastics (Springer Open), Microplastics (MDPI), handbooks, and conferences (MICRO) all dealing particularly with this topic. Numerous research questions are being pursued regarding microplastics, for example concerning their sources, to what extent they are entering the environment, how microplastics behave in the environment (fate), in what manner they interact with flora and fauna, and what effects this entails.

In the contribution by a group of environmental scientists working on different aspects of microplastics, *Maren Heß and colleagues* give an overview of how microplastics can be identified, monitored, analyzed, and assessed regarding the damage they cause for the aquatic environment. It becomes clear that microplastics challenges science in many ways as it is a complex material that needs new methods and new ways of risk assessment. The complexity is already reflected in the search for a definition of microplastics. The term "microplastics" refers to diverse particles with different chemical compositions, shapes, densities, polymer types, surface properties, sizes, and origins. Environmental scientists and citizen science have disclosed the diverse emission pathways and sources that are strongly connected to usage patterns. So far, there is no consensus on the basis of which properties microplastics should be defined. The discussion on definitions can become important when it comes to political regulations. Questions might arise as to where to draw the line, what is covered by microplastics regulations and what is not. Is it size that is mostly relevant? Or what about blend materials which might consist only in minor parts of synthetic polymers? The chapter by Heß et al. also demonstrates that not only the sampling of microplastics (for example in rivers) requires extensive survey designs in order to cover different layers in the water column, but also the particle analysis used to be very time-consuming in the beginning when it was still done manually. Due to different measurements the results of many studies cannot be compared or put in relation to each other. Therefore, there are frequent demands for standardization and harmonization concerning the methods used.

When it comes to questions of adverse effects and harm of microplastics, we are once again faced with the complexity of microplastics. In general, there are different possible effects that are grouped into ecological, mechanical, and toxicological clusters and that can be observed on different scales, for instance regarding the population as a whole or individual, cellular or subcellular levels. Most of the current testing methods that identify effects, cannot do justice to the diversity of microplastics and its diverse possible effects. Therefore, Heß et al. call for a multi-dimensional testing strategy, "which considers substance parameters (e.g., polymer type, size, and shape), single contaminant or mixture toxicity (e.g., known virgin particles vs. unknown mixture of environmental particles), and the choice of the corresponding endpoint [evidence to characterize the effects] and the resulting suitable test method" (Heß et al., 72). The question of "how much information is enough for action" is a recurring point of discussion in environmental sciences (Backhaus and Wagner 2020). In the case of microplastics, "from a purely scientific point of view, each particle could be an individual multiple stressor in itself—requiring an individual testing strategy" (Heß et al., 76). This is, of course, not feasible from a regulatory point of view. The authors are suggesting precautionary measures referring to the persistence of microplastics, their further accumulation in the environment without any action and the difficulty to assess long-term effects on ecosystems. Here, we see that the risks imposed by microplastics reflect characteristics that have been described as typical for environmental risks in general (e.g., uncertainty, complexity, delayed consequences) (Kramm and Völker 2018; Steg and de Groot 2018).

Media coverage on scientific findings and the work of activists have raised public concern about microplastics. Surveys on public risk perception show that for the majority of the public perceives (micro-)plastics have a major risk which is why there is a certain concern (for Europe see Eurobarometer, for Germany see Kramm et al. 2022). In their contribution, *Marcos Felipe-Rodriguez, Gisela Böhm and Rouven Doran* analyze from an environmental psychological perspective which factors contribute to a risk perception of microplastics. While only few studies on risk perception of microplastics exist, they discuss different heuristics and concepts with respect to the topic of microplastics. Drawing on the psychometric paradigm (Fischhoff et al. 1979), they analyze the main risk perception features of microplastics that show characteristics of an "unknown risk", since it is a rather new risk and not easy to observe including its potential delayed effects. Furthermore, they discuss how specific perceiver characteristics, such as socio-demographics, reasoning knowledge and fairness, as well as values and worldviews impact risk perception. Another important factor are emotions, that are grouped according to consequentialist and moral emotions. Microplastics activate both forms of emotions. While consequential emotions comprise fear, for example about consequences of microplastics for sea animals, moral emotions on the other hand are linked to outrage and guilt over violating moral norms, such as the plastic pollution of environments caused by humans. Next to emotions, mental models, that is how people mentally conceive microplastics regarding their causes, consequences, and solutions, are also crucial for risk perception research on microplastics.

The question of how perception and awareness correspond to action and practice is addressed in the next contribution. Based on a survey and group discussion *Immanuel Stieß, Luca Raschewski and Georg Sunderer* explore along the lines of certain practices concerning consumer goods such as peelings and fleece jackets. They consider the awareness of consumers and their knowledge about this set of issues as well as their attitudes regarding alternative practices. It turns out that plastic usage is deeply interwoven with social practices. There are many plastic products used on a daily basis that release microplastics which then enter into our environment either via our wastewater or via the air. The authors investigate practices around peelings, fleece jackets, moist toilet paper, and dog poop bags. They identify promoting and inhibiting factors for the adoption of more environmentally friendly practices and they investigate who the consumers perceive to be responsible to take action. The authors ascertain a high problem awareness and overall concern regarding microplastics on the part of the consumers. However, the latter often lack knowledge of the environmental impacts of their product use. Respondents were for example not aware of the fact that moist toilet papers contain plastic fibers. Thus, consumers do not link their practices to the release of microplastics. Respondents of the survey attributed producers and manufacturers of the products the most responsibility for preventing microplastics to be discharged into wastewaters.

While Stieß, Raschewski and Sunderer look on daily practices of plastic usage, *Marine Isabel Severin, Alexander Hooyberg, Gert Everaert and Ana Isabel Catarino* engage with questions of the psychological aspects of participating in citizen science projects. In their literature review, they analyze what impacts citizen science projects such as public beach clean-ups and plastics surveying activities have on the participants regarding well-being, human health, and ocean literacy. While there is much literature on the scientific outputs of such projects, the role of the participants has so far not been analyzed in-depth. Severin et al. turn to the existing literature in order to identify educational and behavioral impacts that occur after the participation in a citizen science activity such as increased environmental awareness and pro-environmental behavior.

Aspects of health and well-being that were identified in the literature, are categorized as of emotional and cognitive origin. The observed emotional changes were described as feeling more "empowered/encouraged" and "grateful" (Yeo et al. 2015) and as "better mood" and "higher meaningfulness" (Wyles et al. 2017). However, they also identify negative aspects "such as a lower cognitive restoration in response to beach cleaning compared to walking or rock pooling" (Severin et al., 153). Severin et al. make an important critical point in their analysis, when pointing out the diversity bias of the participants and the socio-economic and geographically uneven distribution of access to and benefits of citizen science activities. "Most participants are from middle to high income backgrounds, with access to education, technical skills, resources, and infrastructure that facilitate engagement in citizen science projects" (Severin et al., 155–156). Since environmental datasets based on citizen science could be used to craft policies and measures, the under- and non-representation of certain geographic areas and specific communities may imply their exclusion when prioritization policies to prevent and mitigate plastic pollution (Pandya and Dibner 2018; Pateman et al. 2021). They argue for empowering "local communities in taking part in the development of a successful plastic circular economy, via perception and behavior shifts, and via active participation in decision-making" (Severin et al., 157). Therefore, they recommend, that plastics-related projects involving citizen scientists should broaden their collaborative scopes to include overlooked geographic areas, marginalized communities and underrepresented socio-economic and socio-demographic groups. Furthermore, future surveying programs of plastic pollution involving citizen scientists should consider a collaboration with natural and social science professionals to achieve an in-depth evaluation of the educational and psychological benefits to the participants while also taking social diversity into account.

Raising awareness is a measure which is quickly referred to when it comes for example to the reduction of the plastics consumption. However, research has shown that awareness alone does not automatically imply a behavior change (Heidbreder et al. 2019; van Valkengoed and Steg 2019). In the literature, this is referred to as "awareness-behavior gap" (e.g., Lio and Bai 2014). While measures that are laying the responsibility for the problem and for the solution solely on individuals, have been criticized in social research, e.g., research on consumption pointing towards structural changes in the economy or the society, the individual level is, nevertheless, acknowledged as one among many elements in the plastic assemblage. Thus, as the authors *Maja Grünzner and Sabine Pahl* argue in their contribution, social and environmental psychology approaches focusing on behavior change can contribute to design measures and interventions to reduce plastic consumption. They show that problem awareness and perception are important elements in understanding current beliefs and opinions, but there are still other elements that also need to be considered when it comes to promoting behavior change. Grünzner and Pahl discuss two theoretical models—a prediction model and a stage model and then present practical guidelines for implementing behavior-targeted intervention tools. Amongst others, these tools include cognitive dissonance, goal setting, instructions, rewards, and prompts. The authors show that knowing about negative impacts of plastics alone does not entail a behavior change since habits as well as social and situational factors, like convenience play an important role. The authors conclude that there is no "one fits all" solution for tackling environmental behavior, since "some behaviors have stronger moral components, some are more influenced by the individual's perception of control (how easy or difficult it is doing a behavior) and others are strongly habitualized" (Grünzner and Pahl, 188). They also point out that behavior change should be understood as just one component amongst others when it comes to tackling plastic pollution. It needs "actions from different stakeholders such as governments, businesses, and communities worldwide combining pre- and postconsumption solutions [...]" (Grünzner and Pahl, 189).

When it comes to interventions on the individual level, often the prevention or usage reduction of single-use plastic products (e.g., plastic packaging, plastic cups or tableware) are targeted. Packaging for example produce the largest share of plastics waste (Ritchie and Roser 2018). However, there is an additional reason to opt for less plastic packaging and packaging in general and that is chemicals. The purity and inertness of plastics have been long deconstructed. The public is aware that there may be additives that can leach from the plastic product. However, food contact materials, such as plastics, count as safe. *Jane Muncke and Lisa Zimmermann* show in their contribution that it is safe to assume that there are many plastic materials on the market that are in contact with food and from which untested chemicals get into food. As the authors demonstrate, there are several reasons for that: i) plastics materials that are in contact with food can consist of hundreds to thousands different chemicals. During the production process (polymerization) non-intentionally added substances (NIAS), comprising polymerization impurities, reaction by-products, and break-down products of additives can occur; ii) chemicals intentionally added to food contact materials are tested only one by one. However, "[m]ixtures are a reality for all plastics in direct food contact, with hundreds to thousands of chemicals found to migrate at the same time, many of which are unknown" (Muncke and Zimmermann, 209) as it is the case for non-intentionally added substances (NIAS). Muncke and Zimmermann criticize the virtually uncontrolled human exposure to chemicals. They point towards epigenetic effects, meaning that the DNA itself does not change, but the activity state of genes changes, due to chemical exposure, which can be passed on to generations to come. They call for the moral obligation to decision-makers to protect future generations. "It is unknown whether industry is performing risk assessments for each and every of these substances. What is clear is that regulatory enforcement agencies are neither assessing all chemicals that are authorized for use in food contact plastics nor controlling whether industry is meeting its regulatory requirements—in part, because the required information is not shared with authorities" (Muncke and Zimmermann, 211). This chemical complexity of plastics also poses a challenge to the process of recycling. Due to limited information on the chemicals in plastic products some might be of concern to waste and recycling institutions, a real "re"cycling is hardly possible, thus for most plastics products it is rather a down-cycling that applies. Another challenge are so called "legacy contaminants". These are substances that are now forbidden, but may still be contained in recycled plastics.

While Muncke und Zimmermann point out the challenges of "recycling" and the circularity of plastic products, *Sandra Eckert* explores the role of plastics in European policies and regulation activities on establishing a circular economy from a political sciences perspective. She also pursues the question if the circular economy agenda promotes the transformation of plastics towards more sustainability taking a specific focus on the policy-industry nexus. She acknowledges that a diverse range of European circular economy policies have recently been adopted to respond to issues of plastic pollution like marine litter and microplastics and that the regulatory pressure on the plastic sector has been intensified in the last years. Eckert questions however, if these measure really lead to a transition to more circularity or if they do not, on the contrary, rather stabilize the current linear economy. Drawing on literature on stakeholder involvement in policy frames and discursive business power as well as rule-setting capacity and strategic mobilization of expertise by corporates (e.g., Fuchs 2007; Mikler 2018), she analyses the circular economy agenda of the European Union with regard to the policy-industry nexus. By taking the earlier debate on PVC as an example, she identifies challenges that are currently recurring for the circular economy of plastics. In both cases, the PVC and the current plastic issue, the industry has intervened in policy debates to avoid worst case scenarios like a ban of PVC and has emphasized its recycling ambitions. However, due to recycling constraints like legacy contaminants, recycling is not the panacea of the plastic waste, rather other measures of the circularity debate need more attention like for instance refuse, rethink or reduce. Eckert points out the balancing act of the policy-makers between teaming up with actors from the industry and endorsing voluntary industry pledges on the one hand and dealing with stringent regulatory actions on the other (like the single-use plastics ban). She argues, though, that the circular economy agenda could too easily be brought in line with the status quo of a linear economy which in the case of plastics is mainly fossil-based despite a few bio-based exceptions. The prognosis is that plastic production as a key branch of the petrochemical industry will become the fastest growing sector in terms of oil consumption (International Energy Agency 2021).

Another critical perspective is formulated in the next contribtion. For a few years now, measures to curb plastic pollution have been publicly debated. Among those measures is the use of plastivores, plastic eating organisms. *Kim De Wolff* explores in her contribution biotech visions of plastivores as panacea to plastic pollution. Scientists have discovered up to 30,000 plastic degrading enzymes and emphasize the crucial role those microorganisms could—with the help of bioengineering—play in the global management of the plastic waste (Zrimec et al. 2021). With the abundance of plastivores the earth is adapting to the plastic pollution and this "natural" process can be used in waste management. Kim De Wolff turns to science and technology studies informed by feminist and queer theory to point out that this "natural" process is only a solution if the problem is dealt with in isolation leaving out the entanglements of persistent harms of toxic chemicals and extractive capitalisms.

Inspired by Heather Davis, she argues for a broadening of relationships of responsibility and care beyond the familiar, to include the own community and even transcend the boundaries of species to explore alternative ways of living and to question current modes of living and their social, economic, and political orders.

A world of interconnectedness counters industrial and political strategies of containment and control (e.g., of chemicals in plastics or microplastics) based on assumptions of separation and purity. By recognizing plastivores as our "queer kin" is already an act of resistance to logics of extractivism and control of nature. Kim De Wolff emphasizes not to adopt the narrative of earth's adaptation to plastics pollution as this shifts responsibility from social to natural processes. Furthermore, integrating microbes in the recycling process perpetuates the plastic production of the petro-capitalism. "Plastic-eating solves a problem for petro-capitalism, not the problem of it" (De Wolff, 272). De Wolff urges scholars of the humanities and natural scientists alike to re-think their relationships to petro-capitalism and calls for radical collaborations to challenge our plastic system.

References

Backhaus, Thomas, and Martin Wagner (2020). Microplastics in the environment: Much ado about nothing? A debate. *Global Challenges*, 4 (6), 1900022.

Bensaude Vincent, Bernadette (2013). Plastics, materials and dreams of dematerialization. In Jennifer Gabrys, Gay Hawkins, and Mike Michael (eds.). *Accumulation. The material politics of plastic*, 17–29. Abingdon, Oxon: Routledge.

Bergmann, Melanie, France Collard, Joan Fabres, Geir W. Gabrielsen, Jennifer F. Provencher, Chelsea M. Rochman, Erik van Sebille, and Mine B. Tekman (2022). Plastic pollution in the Arctic. *Nature Reviews Earth & Environment*, 3 (5), 323–337.

Fuchs, Doris (2007). *Business power in global governance*. Boulder: Lynne Rienner.

International Energy Agency (2021). World energy outlook 2021. 03.03.2022 https://www.iea.org/reports/world-energy-outlook-2021.

Haider, Tobias P., Carolin Völker, Johanna Kramm, Katharina Landfester, and Frederik R. Wurm (2019). Plastics of the future? The impact of biodegradable polymers on the environment and on society. *Angewandte Chemie International Edition*, 58 (1), 50–62.

Heidbreder, Lea Marie, Isabella Bablok, Stefan Drews, and Claudia Menzel (2019). Tackling the plastic problem: A review on perceptions, behaviors, and interventions. *Science of the Total Environment*, 668, 1077–1093.

Fischhoff, Baruch, Paul Slovic, and Sarah Lichtenstein (1979). Weighing the risks: Risks: Benefits which risks are acceptable? *Environment: Science and Policy for Sustainable Development*, 21 (4), 17–38.

Kerber, Heide, and Johanna Kramm (2021). On-and offstage: Encountering entangled waste–tourism relations on the Vietnamese Island of Phu Quoc. *The Geographical Journal*, 187, 98–109.

Kerber, Heide, and Johanna Kramm (2022). From laissez-faire to action? Exploring perceptions of plastic pollution and impetus for action. Insights from Phu Quoc Island. *Marine Policy*, 137, 104924.

Kingdon, John W. (2014). *Agendas, alternatives, and public policies*. Harlow: Pearson.

Klinke, Andreas, and Ortwin Renn (2006). Systemic risks as challenge for policy making in risk governance. *Forum Qualitative Sozialforschung/forum: Qualitative Social Research*, 1 (7), Art. 33.

Kramm, Johanna, and Carolin Völker (2018). Understanding the risks of microplastics. A social-ecological risk perspective. In Martin Wagner, and Scott Lambert (eds.). *Freshwater microplastics*, 223–237. Cham: Springer.

Kramm, Johanna, Carolin Völker, Tobias Haider, Heide Kerber, Lukas Sattlegger, Lukas, and Lisa Zimmermann (2020). *Sozial-ökologische Forschung zu Plastik in der Umwelt. Ergebnisse der Forschungsgruppe PlastX*. Frankfurt am Main: ISOE – Institut für sozial-ökologische Forschung.

Kramm, Johanna, Stefanie Steinhoff, Simon Werschmöller, Beate Völker, and Carolin Völker (2022). Explaining risk perception of microplastics: Results from a representative survey in Germany. *Global Environmental Change*, 73, 102485.

Kramm, Johanna (2022). Determining risks of microplastics – Boundary practices of science and authorities. *Environment and Planning E: Nature and Space* (submitted).

Liu, Yong, and Yin Bai (2014). An exploration of firms' awareness and behavior of developing circular economy: An empirical research in China. *Resources, Conservation and Recycling*, 87, 145–152.

Meikle, Jeffrey (1995): *American plastic: A cultural history*. New Brunswick, London: Rutgers University Press.

Mikler, John (2018). *The political power of global corporations*. Cambridge: Polity Press.

Moore, Charles J., Shelly L. Moore, Molly K. Leecaster, and Stephen B. Weisberg (2001). A comparison of plastic and plankton in the North Pacific central gyre. *Marine Pollution Bulletin*, 42 (12), 1297–1300.

Unmüßig, Barbara (2021). The myth of good plastic. 24.06.22 https://www.boell.de/en/2020/11/02/das-maerchen-vom-guten-plastik.

Pandya, Rajul, and Kenne Ann Dibner (eds.). (2018). *Learning through citizen science: Enhancing opportunities by design*. Washington, D.C.: National Academies Press.

Pateman, Rachel, Alison Dyke, and Sarah West (2021). The diversity of participants in environmental citizen science. *Citizen Science: Theory and Practice*, 6 (1), 1–16.

PlasticsEurope (2020). Plastics – the facts 2020. An analysis of European plastics production, demand and waste data. 20.05.2022 https://plasticseurope.org/knowledge-hub/plastics-the-facts-2020/.

Ritchie, Hannah, and Max Roser (2018). Plastic Pollution (Our World in Data). 20.07.2022 https://ourworldindata.org/plastic-pollution.

Sattlegger, Lukas, Immanuel Stieß, Luca Raschewski, and Katharina Reindl (2020). Plastic packaging, food supply, and everyday life. *Nature and Culture*, 15 (2), 146–172.

Sattlegger, Lukas (2021a). Making food manageable – Packaging as a code of practice for work practices at the supermarket. *Journal of Contemporary Ethnography*, 50 (3), 341–367.

Sattlegger, Lukas (2021b). Negotiating attachments to plastic. *Social studies of science*, 51 (6), 820–845.

Sklar, Robert (ed.). (1970). *Plastic Age (1917–1930)*. New York: George Braziller.

Statista (2021). Global plastics industry - statistics & facts. 15.12.2021 www.statista.com/topics/5266/plastics-industry/#dossierKeyfigures.

Steg, Linda, and Judith I. M. de Groot (2018). *Environmental psychology. An introduction*. Second edition. Hoboken, NJ: Wiley-Blackwell; John Wiley & Sons Inc.

Thompson, Richard C., Ylva Olsen, Richard P. Mitchell, Anthony Davis, Steven J. Rowland, Anthony W. G. John, Daniel McGonigle, and Andrea E. Russell (2004). Lost at sea: Where is all the plastic? *Science*, 304 (5672), 838–838.

UNEA (2022). Draft resolution. End plastic pollution. Towards an international legally binding instrument. United Nations Environment Assembly of the United Nations Environment Programme. Nairobi. 20.07.2022 https://wedocs.unep.org/bitstream/handle/20.500.11822/38522/k2200647_-_unep-ea-5-l-23-rev-1_-_advance.pdf?sequence=1&isAllowed=y, .

van Valkengoed, Anne M., and Linda Steg (2019). Meta-analyses of factors motivating climate change adaptation behaviour. *Nature Climate Change*, 9 (2), 158–163.

Völker, Carolin, Johanna Kramm, and Martin Wagner (2020). On the creation of risk: Framing of microplastics risks in science and media. *Global Challenges*, 4 (6), 1900010.

WEF (2016). The new plastics economy. Rethinking the future of plastics, 2016. 15.09.2016 http://www3.weforum.org/docs/WEF_The_New_Plastics_ Economy.pdf.

Wyles, Kayleigh J., Sabine Pahl, Matthew Holland, and Richard C. Thompson (2017). Can beach cleans do more than clean-up litter? Comparing beach cleans to other coastal activities. *Environment and Behavior*, 49 (5), 509–35.

Yeo, Bee Geok, Hideshige Takada, Heidi Taylor, Maki Ito, Junki Hosoda, Mayumi Allinson, Sharnie Connell, Laura Greaves, and John McGrath (2015). POPs monitoring in Australia and New Zealand using plastic resin pellets, and International Pellet Watch as a tool for education and raising public awareness on plastic debris and POPs. *Marine Pollution Bulletin*, 101 (1), 137–45.

Zimmermann, Lisa, Georg Dierkes, Thomas A. Ternes, Carolin Völker, and Martin Wagner (2019). Benchmarking the in vitro toxicity and chemical composition of plastic consumer products. *Environmental Science & Technology*, 53 (19), 11467–11477.

Zimmermann, Lisa, Andrea Dombrowski, Carolin Völker, and Martin Wagner (2020a). Are bioplastics and plant-based materials safer than conventional plastics? In vitro toxicity and chemical composition. *Environment International*, 145, 106066.

Zimmermann, Lisa, Sarah Göttlich, Jörg Oehlmann, Martin Wagner, and Carolin Völker (2020b). What are the drivers of microplastic toxicity? Comparing the toxicity of plastic chemicals and particles to Daphnia magna. *Environmental Pollution*, 267, 115392.

Zimmermann, Lisa, Zdenka Bartosova, Katharina Braun, Jörg Oehlmann, Carolin Völker, and Martin Wagner (2021): Plastic products leach chemicals that induce in vitro toxicity under realistic use conditions. *Environmental Science & Technology*, 55 (17), 11814–11823.

Zrimec, Jan, Mariia Kokina, Sara Jonasson, Francisco Zorrilla, and Aleksej Zelezniaka (2021). Plastics and the microbiome: Impacts and solutions. *Environmental Microbiome*, 16 (2), 1–19.

Explaining Agenda-Setting of the European Plastics Strategy. A Multiple Streams Analysis

Paula Florides, Johanna Kramm

Introduction

Since the start of plastics mass manufacturing in the 1950s, plastic production volumes have grown continuously, reaching almost 370 million tonnes globally in 2020 (PlasticsEurope 2021, 12). The public image of plastics, simultaneously, has changed for the worse: In a representative survey conducted in 2017, 87 percent of Europeans expressed worry about the effects of plastics on the environment and 74 percent about its effects on their health (European Commission 2017). Politically, plastics have become an integral part of European environmental discussions and in 2018, concerns about growing plastic waste pollution were met by the EU with the adoption of the European Strategy for Plastics in a Circular Economy. As the first EU-wide material-specific policy framework, the strategy aims to lay the "foundations to a new plastics economy" (European Commission 2018, 1) by targeting the entire life-cycle of plastics from design, use, reuse, to recycling (for a critical discussion on the circularity of plastics see Eckert in this volume). Against the backdrop of plastics being "so prominent on the political and societal agenda that the issue could become a policy field of its own" (Mederake and Knoblauch 2019, 11) the narrative of plastics as an environmental pollutant is however by no means new. Instead, newspaper reports publicly drawing attention to the issue can be traced back to the 1970s (Kramm and Völker 2018, 225). Therefore, this contribution aims to understand how and why plastics shifted from being a long-known environmental problem, only sporadically addressed in European legislation to one that, in recent years, has gained such high political salience and was placed on the top of the EU's agenda. To do so, the contribution addresses the question: How did agenda-setting take place in the case of the European Strategy for Plastics in a Circular Economy?

The contribution is structured around the Multiple Streams Framework (MSF), which offers a comprehensive account of the agenda-setting process by incorporating human agency, temporal order as well as contextual and institutional factors. Since its introduction in the 1980s by John Kingdon, it has shaped the course of agenda-setting studies fundamentally and to this day remains one of the most frequently cited frameworks on agenda-setting (Herweg 2013; Zahariadis 2016). Regarding the European plastics strategy, the MSF proves particularly useful as it takes into account factors such as institutional fluidity, jurisdictional overlap, endemic political conflict, varying time cycles and policy entrepreneurship, which are typically considered to be "pathologies of the EU system" (Ackrill et al. 2013, 871). After an introduction to relevant MSF terminology and remarks on the methods applied, the contribution provides a presentation and discussion of the results of the EU's agenda-setting process on plastics.

The Multiple Streams Framework

The MSF aims to explain the process of agenda-setting under conditions of ambiguity (for a graphic overview see Fig. 1). Ambiguity, here, refers to the constant fluctuation of participants within institutions where a different set of actors is involved in each step of the policy process, and preferences typically differ, meaning that the same problem is interpreted in disparate ways by different actors (Zahariadis 2007, 66 f.). In such an ambiguous environment, gathering information is a difficult task for policy-makers, according to the MSF, as they are bound by considerable time constraints, and must, inevitably, make decisions with limited information at hand. This paves the way for political manipulation as information can be provided strategically to "alter the dynamics of choice by highlighting one dimension of the problem over others" (Zahariadis 2007, 70). Accordingly, the key actors behind agenda change are policy entrepreneurs (PEs) who actively exploit politically opportune moments to influence policy outcomes. PEs can be individuals or groups inside or outside of the government (e.g., politicians, business persons, citizens, NGOs...) (Frisch Aviram et al. 2020, 619 ff.). Similarly, their reasons for getting involved are versatile and can range from practical and material incentives where a direct, personal, concrete gain is at stake to ideological motives (Kingdon 2014, 123) and a variety of different PE strategies have been identified in agenda-setting literature (Frisch Aviram et al. 2020, 625 ff.). All PEs, however, share the defining characteristics of their willingness to devote a considerable amount of resources (e.g., time, energy, reputation, money) in the hope of a future return (Kingdon 2014, 179) as well as continuous efforts in "softening up" relevant stakeholders to build acceptance for their policy proposals (Kingdon 2014, 128). Nonetheless, PE activities alone do not suffice in explaining agenda-setting, which is fundamentally shaped by the context regulating social circumstances and interactions (Zahariadis and Exadaktylos 2016). This account of context is provided in the MSF through the concept of three separate streams and ripeness in all three must first be achieved to enable agenda-change: In the problem stream, attention is drawn to a problem through indicators (routine monitoring or studies conducted by government agencies, non-governmental researchers, or academics), focusing events (sudden events such as crises or disasters providing a powerful impetus for policy change), or feedback from previous programs (Kingdon 2014, 90 ff.). Problem stream ripeness is achieved "if, first, one of the above-mentioned mechanisms raises attention to an issue, and, second, that issue is interpreted as being problematic" (Herweg 2016, 18) by the respective policy-makers. Such problem recognition can be influenced by PEs or problem brokers who both engage in problem framing and promoting issues to policy-makers. However, unlike PEs, problem brokers refrain from making suggestions on how to solve the problem beyond the notion that something needs to be done (Knaggård 2015, 453). In the policy stream, solutions in the form of different policy proposals compete for consideration by policy-makers. A proposal's survival is dependent on its value acceptability (its conformity to existing value constraints), technical feasibility (the technical possibility of implementing the proposal) and resource adequacy (obtainability of the needed resources) (Jones et al. 2016, 16). Ripeness in the policy stream is achieved when a policy proposal adhering to these prerequisites presents itself and is successfully promoted to policy-makers.[1] The politics stream comprises the institutional and

---

[1] It has been argued that since policy communities in the EU include national, European as well as transnational members they are "inevitably characterized by considerable heterogeneity" making agreeance upon acceptability, tolerable costs, normative acceptance, and receptivity among decision-makers highly unlikely and that, therefore, stream ripeness may be reduced to the criterion of technical feasibility in EU-level MSF applications (Herweg 2016, 21).

cultural context of the output of concern and consists of three main elements which each influence the likelihood of policy-makers promoting certain items on the agenda: Public mood (the perception of the public and other relevant stakeholders thinking along similar lines), pressure-group campaigns (the support or opposition of interest groups), and personnel changes in the political-administrative system itself (e.g., due to election results or changes in ideological distributions which may result in a cyclical upturn of certain issues). Stream ripeness is achieved by the presence of either of the three. According to the MSF, each of the streams has a life of its own and develops and operates largely independently and unrelated to the others (Kingdon 2014, 85). However, during critical moments in time, a fleeting window of opportunity opens when the attention of policy-makers is drawn to a specific policy problem or solution due to one of the before mentioned mechanisms. An issue ascends onto the agenda during such opportune moments if a PE succeeds in coupling the three streams by linking a problem and their preferred solution and successfully promoting it to the right actors at the right point in time. An issue is considered to have entered the agenda when it is not only paid serious attention to by policy-makers but also moved into position for an active decision (Kingdon 2014, 166).

Explaining Agenda-Setting

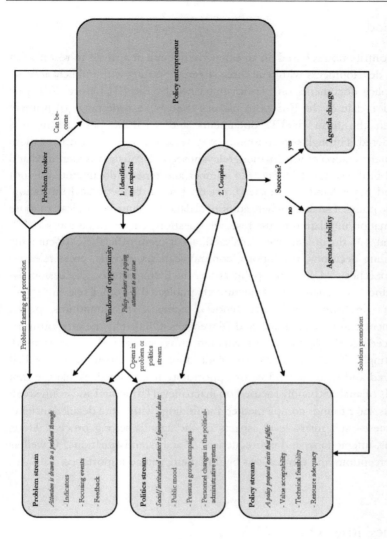

Fig. 1: *Multiple Streams Framework. (Source: own graphic)*

Methods

The contribution is based on an extensive document analysis to reconstruct major developments within all three streams. Starting with publicly accessible policy documents on plastics published by the EU before 2018 (the adoption date of the European plastics strategy), a wide range of primary data in the form of EU documents and scientific publications was considered. Through several rounds of systematic extraction, synthetization, and interpretation of information, relevant secondary data was identified and added in the process. Hence, the number and type of documents was not defined beforehand, but, instead, oriented towards theoretical saturation, i.e., the point of analysis, where additional data sampling did not lead to more information in relation to the research question (Glaser and Strauss 2017). In total, 131 documents were analyzed; 49 of them official EU documents (including decisions, resolutions, communications, reports, press releases, and speeches), 44 scientific publications (including articles, discussion papers, and book chapters), 24 documents published by other relevant stakeholders (including conference reports, technical memorandums, public statements, and declarations), and 14 websites. Guided by the previously introduced MSF features, information was deductively systematized. Resultingly, inferences could be made about how stream ripeness had occurred and when a window of opportunity had opened. While the first rounds of analysis mainly focused on macro-level (structural and contextual) aspects, later rounds complemented the findings with more detailed insights into meso- and micro-level aspects of the agenda-setting process. Here, actions, interactions, and consequences of actors and situations, as well as the perceptions and motivations behind them, gained importance.

Plastics' Rise onto the EU's Agenda

Plastics have a history of being implicitly covered within two separate policy debates: In the context of marine litter, they have played a role from the onset of the discourse in the 1970s when indicators increasingly appeared in scientific literature demonstrating the ubiquity of plastic particles at sea (e.g., Carpenter et al. 1972; Carpenter and Smith 1972; Colton et al. 1974; Wong et al. 1974). Plastics were problematized in different contexts from concerns on a rise in mortality of different species of seabirds, seals, fish, turtles, and other marine mammals due to the ingestion of different types of plastic materials or the entanglement therein (e.g., Kenyon and Kridler 1969; Parslow and Jefferies 1972; Rothstein 1973; Cornelius 1975; Baltz and Morejohn 1976; Day et al. 1985; Harper and Fowler 1987) to beach litter (e.g., Cundell 1973; Dixon and Cooke 1977; Gregory 1978), or as means of altering ecosystems by introducing non-native species (e.g., Winston 1982).

The same problem recognition, however, was not reflected in the policy instruments that were adopted as a response to the growing concerns about ocean pollution at that time. In the plethora of different global, regional, national and local conventions, agreements, action plans, and initiatives, plastics were only implicitly covered within the broader context of marine litter[2], and, at this, exclusively maritime sources of pollution were subject to regulation (Gold et al. 2013; Simon and Schulte 2017; Simon et al. 2018). In the EU, marine litter was considered the responsibility of coastal member states and the respective Regional Seas Conventions (European Commission 2011b, 16) and was not explicitly connected to plastics until the early 2010s when two workshops and an in-depth report on the issue of marine plastic debris were issued (European Commission 2010b; European Commission and DG ENV 2011; European Parliament 2013). Similar is the case in European waste legislation. Highlighted as a priority as early as in the First Environmental Action Plan, adopted in 1972, waste management "has continued as a priority area for EU action with a number of important Commission policy dossiers having been published to determine and evaluate the direction of EU waste law" (Farmer 2012, 3). However, until the publication of a Green Paper on a European Strategy on Plastic Waste in the Environment (European Commission 2013b) by the European Commission in 2013, plastics were mentioned in discussions on EU waste policy only in the context of general recycling targets for packaging and household waste or as individual product groups (e.g., phthalates and PVC). Based on these observations, a window of opportunity for agenda change can be assumed to have opened in the early 2010s. A closer look at developments within

---

2 An exception to this is MARPOL Annex V on the Prevention of Pollution by Garbage from Ships, which entered into force in December 1988, and (besides specifying the distances from land and the manner in which different types of garbage may be disposed of) includes a complete ban on the dumping of operational plastic waste generated on ships into the sea.

problem, policy, and politics stream reveals how the window manifested and how stream coupling took place enabling a shift onto the EU's agenda.

Problem and Politics Stream

As can be observed in this case, the existence of indicators merely showing that the problem is "out there" was not enough to achieve a European problem recognition. To qualify as a problem in the MSF's sense and to achieve ripeness in the problem stream, the condition additionally needed to be framed in a way that required action by policy-makers (Herweg 2016, 18). This entailed (1) an explanation why something must be done about the issue and (2) a clarification why it is the EU that should be doing something (Princen 2009, 42). Regarding both premises, several developments within the problem stream can be assumed to have made an impact:

A Question of Why: Creating Salience

Problem Framing

After most of the threats of marine litter to marine systems had been identified throughout the 1970s and 80s, research on the topic declined throughout the 1990s and was largely restricted to monitoring trends assessing the effectiveness of mitigation measures until a new wave of research interest was stimulated in the early 2000s (Ryan 2015). This resurge in interest can largely be attributed to a publication by Moore et al. in 2001 on an alarmingly high accumulation zone of plastics in the North Pacific Ocean. It notoriously became known as the "Great Pacific Garbage Patch" (GPGP) or "plastic soup" creating new problem terminology that was spread by mass media. (Council of the European Union 2009; European Commission 2010a; 2010b; Potočnik 2010; European Commission and DG ENV 2011). Similarly, although knowledge of the degradation of plastic debris into ubiquitous and widely dispersed small particles can be traced back to its recognition as a significant marine pollutant in the 1970s, it was only the introduction of the term "microplastics" (first coined by Thompson et al. in 2004 and popularized by Eriksen et al. in 2014) as a label for such particles (previously described as pellets, fragments, spherules, granules, etc.)

which "propelled their scientific career" (Kramm and Völker 2018, 227) leading to an exponentially growing number of studies discovering microplastics in more and more ecosystems and producing knowledge on its sources, environmental fate, and biological effects (see Fig. 2).

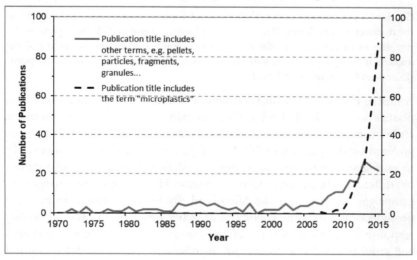

Fig. 2: Environmental studies on plastic particles from 1970 to today. (Source: Kramm and Völker 2018, 228)

A range of scientific publications in the 2000s further highlighted the special role of plastics in marine pollution (e.g., Barnes 2002; Derraik 2002; Barnes et al. 2009; Gregory 2009; Ryan et al. 2009). Such publications played an important role in problem recognition and in setting the terminology with which plastics would subsequently be addressed in policy discussions on marine litter (e.g., Council of the European Union 2009; European Commission 2010a; 2010b; Potočnik 2010; European Commission and DG ENV 2011). Here, the role of researchers and activists oftentimes overlapped as many researchers actively advocated for the issue. Their findings were not only "cited in the majority of the policy papers on plastic marine litter" but researchers were also "present at most of the stakeholder events partly using emotionalizing strategies to create urgency" (Langer 2017, 53). A prominent example is activist and oceanographer Captain Charles Moore who, after discovering the GPGP, continued to make it his "life's work" (Moore, no date) to raise awareness on the issue of plastic marine pollution. By approaching both, politicians as well as the general public on several occasions from the early 2000s onwards[3], Moore actively engaged in problem promotion and, in 2010, reached the EU directly at the European Commission's workshop on plastic marine litter, as noted in the workshop's protocol:

> Charles Moore, founder of the Algalita Marine Research Foundation in the USA and the discoverer of the Great Pacific Garbage Patch, illustrated with many pictures the pressure of marine litter in the oceans, the estimated quantities involved and the impacts it has. [...] He ends with the question: How big is your plastic footprint? (European Commission 2010b, 1)

A further example is scientist and activist Marcus Eriksen. After researching plastics in the North Pacific Gyre as part of the Algalita Foundation, Eriksen, together with fellow activist Anna Cummins, founded the nongovernmental organization (NGO) 5 Gyres which is dedicated specifically to raising awareness on marine plastic pollution by organizing expeditions to different plastic accumulation zones. He furthermore engaged in campaigns such as Beat the Microbead. By being persistent in problem promotion and stressing conditions as public problems in need of a policy response by decision-makers, different non-state actors, such as scientists and activists, acted as influential problem brokers (Knaggård 2015, 453) and contributed significantly to directing policy-makers' attention towards plastics. Due to the nature of the problem, which is heavily dependent on scientific knowledge in assessing the magnitude, sources, distribution, and impacts of plastic pollution, specialist knowledge on the issue was required and strengthened the agency and problem framing power of such problem brokers. The same sector-specific framing power, however, may also serve as a resource for industry actors to build counter narratives (see Eckert in this volume).

Pressure Group Campaigns, Mass Media, and Public Mood

Early campaigns against marine litter had predominantly consisted of clean-up operations performed by local authorities, volunteers, and NGOs (United Nations Environment Programme 2005). While these campaigns made an essential contribution to problem recognition by assessing the

---

3 Moore was interviewed, for instance, in popular TV formats such as *The Oprah Show*, the *Late Show with David Letterman*, and *Good Morning America*, published academic as well as non-academic articles and held a *Ted-Talk* in the late 2000s publicly raising awareness on the topic.

magnitude of the problem and educating the public about it (Hidalgo-Ruz and Thiel 2015), the high tide of activism targeting marine litter pollution, and plastics in particular, however, only truly started in the early 2010s.[4] As delineated previously, for the creation of salience, the "facts" of an issue do not necessarily have to change, but the meaning attributed to those facts (Jones 1994, 106). This can be observed in the case of microplastics when, in the early 2010s, a range of high-profile pressure group campaigns demanded a complete ban of microbeads in cosmetic products. Campaigning methods included, for instance, shopping guides listing producers using microplastics in their products or the launch of the app Beat the Microbead in 2012, which could be used to check whether a product contains plastics (Kramm and Völker 2018, 231). This, in addition to concerns about environmental impacts, connected the discussions on microplastics to consumer health and led to a range of national interventions with the first ban of microbeads in cosmetic products in Europe enacted by the Netherlands in 2014. Such focus on consumer health had already been successful on EU-level in the 1990s when targeted NGO campaigns against PVC and phthalates had led to an emergency ban on phthalates in toys in 1999, eventually transformed into a permanent ban in 2005 (Eckert 2019; Eckert in this volume). Salience was furthermore created by mass media: Through newspaper articles and movies, such as Plastic Planet (2009), Trashed (2012) or, later, the BBC series Blue Planet II (2017), labels (e.g., "GPGP", "plastic soup") and powerful symbols (e.g., images of marine organism entanglement/ingestion, beach litter), were created that made problem frames more accessible to broader publics and emotionalized the problem softening up, both, the public and politicians.[5]

Pressure group campaigns and the media had a significant impact on the public mood: A survey commissioned by the Commission and the

---

4 A majority of the prominent organizations, campaigns, and coalitions actively campaigning against plastic marine pollution were only founded in 2009 or later: E.g., 5 Gyres Institute (founded in 2009); Plastic Pollution Coalition (initiated in 2009 by Earth Island Institute); Plastic Soup Foundation (founded in 2011 and initiated the Beat the Microbead campaign in 2012); Trash Free Seas Alliance (initiated in 2012 by Ocean Conservancy); The Ocean Cleanup (founded in 2013); #breakfreefromplastic Movement (founded in 2016); Rethink Plastic Alliance (founded in 2017; alliance of leading European NGOs)

5 The trailer of the movie Trashed (2012) reached European policy-makers directly when it was screened at the International Conference on Prevention and Management of Marine Litter in European Seas in 2013 (European Commission et al. 2013, S. 1).

Directorate-General for Environment (DG ENV) found that by December 2013, nine out of ten Europeans were in favor of greater action on marine litter (European Commission 2014, 17) and, by 2017, 87 percent of Europeans expressed worry about the effects of plastics on the environment and 74 percent about effects on their health (European Commission 2017). The negative public image of plastics is furthermore reflected in the global "anti-plastic bag norm" (Clapp and Swanston 2009) that spread among member states throughout the 2000s where single-use plastic carrier bags were increasingly perceived as "symbolic of a 'throw-away' society" (BIO Intelligence Service 2011, 23). Most evidently, a "climate of the times" is revealed through the reactions of high-stake industry actors: On producer end, several companies—among them global players such as Johnson & Johnson and Unilever—responded to the growing pressure with voluntary action, for instance by pledging to remove microbeads from personal care products (Kramm and Völker 2018, 231). The global plastics industry, too, expressed concerns about the negative image of plastics (European Commission 2010a, 1) and reacted by taking on a proactive role through sponsoring a range of activities and forming global partnerships to contribute to solving the marine litter problem (Eckert 2019, this volume). The notion "that important policy-makers, opinion leaders, and other politicos think along similar lines" (Zahariadis 2008, 518) created pressure on European policy-makers:

> We don't have a lot of time—not least because the European Parliamentarians and the NGOs are getting restless. They/you want action from all of us. Europeans, our seas, our beaches, our ecosystems and our marine-dependent economies deserve nothing less. (Potočnik 2010, 3)

Later statements regarding the adoption of the European plastics strategy emphasized the same sense of urgency. In 2018, then First Vice-President of the Commission, Frans Timmermans, for instance, underlined that "[t]here is a huge sense of urgency in European society that this needs to be done and it needs to be done now" (Timmermans, cited in Mederake and Knoblauch 2019). Similarly, in 2019, then Commissioner for environment, maritime affairs and fisheries, Karmenu Vella called it "one of the most called for and supported EU initiatives among European citizens" (Vella, cited in Mederake and Knoblauch 2019). This observation of a public mood is highly interesting as, in MSF literature, there is disagreement over the existence of a European public mood and whether it should be included in EU-level MSF applications (e.g., Zahariadis 2008; Herweg 2016). As shown, in this case it is useful to include the mood of the European public in the analysis.

Together, the efforts of problem brokers, as well as pressure group and media involvement, created symbols that served as a "reinforcement for something already taking place" (Kingdon 2014, 97) powerfully focusing attention towards plastics. From the 2000s onwards, the material started to stand out from other waste fractions and marine litter became an "obligatory passage point" (Langer 2017, 47) in policy debates on plastics (European Commission 2010b; 2012a; Potočnik, 2010, 2011). This new understanding of the issue is reflected in EU discourses on the topic after 2010, where the terms "plastics" and "marine litter" were oftentimes used interchangeably (European Commission 2010b; 2012a; Potočnik, 2010, 2011):

If we want to win our fight against marine litter we have to know more about it. This is why we will promote more research on the impacts of micro-plastics, the toxicity of plastics and the potential of bio-degradable plastics. (Potočnik 2011, 5)

Resultingly, ripeness in both problem and politics stream was achieved as plastics were now perceived as a distinct issue in need of policy response. As opposed to the MSF's assumption that the streams will remain largely independent until a window of opportunity opens, these findings reveal a close interrelatedness of developments within the problem and politics stream: It could be argued that without the urgency created by pressure group campaigns, and the media in the politics stream indicators might not have reached the attention of policy-makers in the first place. Simultaneously, the indicators provided in the problem stream served as a prerequisite for the formation of many of the pressure group campaigns and media reports.

A Question of Who: "Europeanization" of the Problem

Although at this stage the issue may have become a European debate, it is not automatically also an EU debate. For this to happen, "policy-makers within the EU need not only to be receptive to the issue, but also willing and able to pick it up" (Princen 2009, 152)—a sense of responsibility must be created. One important development in this was a shift in the discourse on marine litter from sea-based to land-based sources from the 1990s onwards (Ryan 2015, 14). This facilitated an EU policy response as it lifted the debate out of the realm of marine policy and the respective patchwork of already existent national, regional, national, and local conventions. As plastic marine litter was increasingly interpreted as a failure of waste management systems, the debate migrated to waste policy—a "particularly active" (European Commission 2015b, 10) area of EU policy. The focus on land-based sources furthermore coincided with findings of high plastic accumulations in European waters, which expanded the discourse on marine litter from the "hot spot" (Ryan 2015, 8) in the Pacific Ocean to a European one:

> Marine litter is a big, big problem. [...] And of course it is not only in the Pacific Ocean. It is a problem in all our seas and oceans. We have all heard about the Great Pacific Garbage Patch which is perhaps the biggest and the most infamous. But you can find the same plastic mess in the seas bordering Europe—and on Europe's coasts. We have to step up with our fight against waste and littering on land in the EU and internationally. (Potočnik 2010, 2)

Attention was "notably drawn to [the] issue" (European Commission 2010a) by the Dutch delegation at the European Council of Ministers for Environment in 2009 and, soon after, by a report drafted by the European Parliament's Environment Committee rapporteur Anna Rosbach both raising concerns on the "plastic soup" now also present in the Atlantic Ocean (Council of the European Union 2009; European Parliament 2010). In response, two workshops and an in-depth report on the issue of marine plastic litter were issued between 2010 and 2013 (European Commission 2010b; European Commission and DG ENV 2011; European Parliament 2013). Additionally, concerns on single-use plastic carrier bags were raised to the EU directly in 2011 after several member states had already adopted national measures, indicating "that effective EU action is needed" (European Commission 2011a, 1). A consultation and a report on options to reduce the use of single-use plastic carrier bags followed in the same year (BIO Intelligence Service 2011; European Commission 2011a). Moreover, from the early 2010s onwards, plastics in the context of marine litter were addressed by a range of international actors, for instance at the UN Rio+20 and GESAMP summits in 2012 or the G7 summit in 2015. This international problem recognition was acknowledged by EU actors (European Commission 2012a; 2012b; 2013b; European Commission et al. 2013a; European Commission 2015a). Due to the newly formed international governance arrangements and commitments the EU could now be held accountable by NGOs and other actors, which increased the pressure to act (Langer 2017, 54). Thus, for problem recognition in the EU to occur, two major important shifts in the way the problem was framed were necessary: Firstly, the shift in the discourse on marine plastic litter to land-based sources, which enabled a connection of the issue to waste policy. This had to be combined with, secondly, findings of plastic debris along European coasts which, through spatial proximity, created a greater sense of EU responsibility. In addition to problem brokers, pressure groups and the media, plastics were furthermore brought to the EU's attention through their thematization at international summits and by individuals and member states who addressed concerns on different aspects of the problem to the EU directly. This interplay of contextual factors and actors opened a window of opportunity in both problem and politics stream in the early 2010s as EU policy-makers were now paying serious attention to the issue and perceived it as being problematic and in need of a policy response.

Institutional Changes, Policy Stream, and Coupling Process

Two institutional changes occurred within the Commission coinciding with the developments in problem and politics stream: In 2010, the second Barroso Commission took office, and Janez Potočnik assumed the role of Commissioner of the DG ENV. Simultaneously, since 2010, climate change, previously a chief policy-making issue under the area of competence of the DG ENV, was now managed by the newly created DG Climate Action. This resulted in a "vacuum" for Potočnik as new Commissioner, which was filled by making "resource efficiency" a major policy priority of the Commission and advancing it as one of the seven flagship initiatives of Europe 2020 (Happaerts 2014, 33). Against the backdrop of the 2008 European debt crisis, resource efficiency was promoted by the Commission as an economic opportunity:

Transforming the economy onto a resource-efficient path will bring increased competitiveness and new sources of growth and jobs through cost savings from improved efficiency, commercialisation of innovations and better management of resources over their whole life cycle. (European Commission 2011b, 4)

The predecessor of Europe 2020, the Lisbon Strategy, had received criticism inter alia due to its lack of attention to social and environmental issues. Therefore, an increased effort was made for Europe 2020 to place such concerns more prominently within the strategy (Happaerts 2014, 32).

Promoting resource efficiency, thus, created a "win-win narrative" (Leipold 2021, 1053) claiming to solve both, environmental and economic issues. Accordingly, the prevailing circumstances within Europe and the Commission itself provided an environment where it was favorable for Potočnik to couple the highly salient problem of plastics to his pet policy proposal of resource efficiency. This connection is reflected in Potočnik's speeches on marine litter between 2011 and 2013:

Once waste is seen as a resource, the demand for it allows for the system to work on its own. Once waste acquires its own market; people want it so that they can use it; to bury or burn it would be rather like burying or burning money. (Potočnik 2011, 4)

I have always seen waste management as part of the wider resource efficiency agenda. I would argue that if we see waste as a resource, it ceases to be a problem; it becomes valuable - people want it. So if litter is "waste that is in the wrong place", then we can follow the same logic and see it as a valuable resource currently in the wrong place. (Potočnik 2013, 3)

Managing plastic waste is a major challenge in terms of environmental protection, but it's also a huge opportunity for resource efficiency. In a circular economy where high recycling rates offer solutions to material scarcity, I believe plastic has a future. I invite all stakeholders to participate in this process of reflection on how to make plastic part of the solution rather than the problem. (Potočnik, cited in European Commission 2013a, 1)

Potočnik, here, acted as a successful PE by coupling problem and solution during an opportune moment in time. The DG ENV engaged in coalition building (Leipold 2021, 1053) and actively promoted this problem-solution package among different venues at a range of different stakeholder events (Potočnik 2011; 2012; European Commission 2013a; Potočnik 2013). The Green Paper on a European Strategy on Plastic Waste in the Environment, published by the Commission in 2013, adopted this focus on resource efficiency and solidified the problem-solution coupling by scheduling a review of policy options to tackle the problem of plastic litter as a part of the wider review of waste legislation in 2014 (European Commission 2013b). Integrating plastics into the pending review of waste policy created ripeness in the policy stream as it offered a technically feasible policy solution. The Green Paper recognized that plastic waste was not specifically addressed by EU legislation "despite its growing environmental impact" (European Commission 2013b, 6)—an assessment that the European Parliament concurred with in a resolution in January 2014 (European Parliament 2014).

Hereby, a European responsibility for managing plastic pollution was officially assumed and plastics moved onto the EU's agenda as the issue was now "up for an active decision" (Kingdon 2014, 4) by policy-makers. Furthermore, the Green Paper set the terms in which solutions to plastic litter were subsequently addressed by the EU's legislative organs:

[The European Parliament] agrees that plastic waste should be treated as a valuable resource by promoting its reuse, recycling, and recovery and by enabling the creation of an adequate market environment [...]. (European Parliament 2014, 3)

The current consumption levels of plastic carrier bags result in high levels of littering and an inefficient use of resources, and are expected to increase if no action is taken. (European Parliament and Council of the European Union 2015, 1)

In the 2015 EU Action Plan for the Circular Economy, plastics occupied the prominent position of the first of five key waste streams that were to be addressed to help stimulate Europe's transition towards a circular economy (European Commission 2015a). The plan, furthermore, announced an explicit strategy on plastics "addressing issues such as recyclability, biodegradability, the presence of hazardous substances of concern in certain plastics, and marine litter" (European Commission 2015a, 14) which was implemented in January 2018 in the form of the European Strategy for Plastics in a Circular Economy (European Commission 2018). The observed entrepreneurial activities by then DG ENV Commissioner Potočnik resonate with findings of other agenda-setting scholars on the Commission's special role in the EU's agenda-setting process (e.g., Cram 1994; Schmidt 2001; Ackrill et al. 2013; Iusmen 2013; Maltby 2013; Menz 2013; Copeland and James 2014; Herweg 2016; Schön-Quinlivan and Scipioni 2017; Brandão and Camisão 2021). Indeed, due to its formal monopoly position to initiate legislation, the Commission was able to act as a "purposeful opportunist" (Cram 1994, 16) "selling" a ready-made package of problem and solution to the other legislative organs. Additionally, the beginning of its term of office opened another window of opportunity as it led to a shift in receptivity to the issue. Nevertheless, the Commission did not act coherently as a single body, but instead, as can be observed in this case, motivations, and actions of individuals within the Commission shaped its entrepreneurial activities significantly.

Conclusion

Overall, the MSF proved highly eligible for the analysis of EU agenda-setting. The analysis of the agenda-setting process within this case revealed, however, a close interrelatedness of developments within the problem and politics stream, contrary to the MSF's notion that, until a window of opportunity opens, streams remain largely independent from one another. Indicators on plastic accumulations and its negative effects in marine environments provided and promoted by non-state actors, most prominently scientists, highlighted the special role of plastics within the broader topic of marine litter and were picked up and popularized by a range of successful pressure group campaigns and mass media. The introduction of new problem terminology (e.g., "GPGP", "plastic soup", "microplastics"), powerful symbols (e.g., images of marine organism entanglement/ingestion and beach litter) and a new risk perception of microplastics in relation to consumer health created salience. The increasingly negative image of plastics was reflected in the public mood and reactions of high-stake industry actors as well as individuals and member states who raised concerns to the EU directly. Together with a heightened sense of responsibility after the shift towards land-based sources and findings of plastic litter along European coasts, as well as international problem recognition, these developments led to ripeness in both problem and politics stream, resulting in the opening of a window of opportunity in the early 2010s. An important role in exploiting the open window and shifting the issue onto the EU's agenda was played by the then DG ENV and its Commissioner Janez Potočnik who actively promoted the problem-solution coupling of plastic waste and resource efficiency, thus achieving policy stream ripeness and facilitating the shift of plastics onto the EU's agenda with the Green Paper in 2013. This observation reinforces previous findings on the European Commission's impact in agenda-setting which can simultaneously be involved in problem framing and promotion, in creating stream ripeness as well as the opening of a window of opportunity. Therefore, the suggestion is put forward to consider both, the Commission's special role as well as the European mood in theoretical considerations on EU agenda-setting.

Agenda-setting, as shown in this case, is the outcome of an interplay of many different complex external factors and human agency. Developments and debates take place on different levels simultaneously and mutually affect each other. Although many critical influencing factors could be identified, this contribution does not claim conclusiveness as preconditions can never be known in their entirety (Kingdon 2014, 73). The results could be reinforced through the addition of further primary data—a step that was not possible within the capacities of this contribution but could be complemented by further research on the topic. Furthermore, in terms of the investigated case, there are several open ends that seem promising for further analysis: Throughout the 2000s, the most prevalent problem frames targeted plastics in the context of marine litter and single-use plastic carrier bags. As delineated previously, however, microplastics in consumer products started to emerge as another highly salient issue in the early 2010s and numerous new campaigns against plastic litter emerged simultaneously. Moreover, relevant institutional changes within the Commission occurred in 2014 when the new Juncker Commission took office. Both events indicate a change in the composition of actors after the stage of agenda-setting. Further analysis could follow up on this lead and inspect how PE strategies and constellations changed as different stakeholders got involved and how this influenced subsequent events before and after the adoption of the European plastics strategy in 2018. This could provide interesting insights not only for the study of EU plastics policy but also for prospective theory development of EU agenda-setting processes.

Author contributions

Paula Florides conceptualized the outline and wrote the manuscript. Johanna Kramm supervised and edited the work.

References

Ackrill, Robert, Adrian Kay, and Nikolaos Zahariadis (2013). Ambiguity, multiple streams, and EU Policy. *Journal of European Public Policy*, 20 (6), 871–887.
Baltz, Donald, and Gentillet V. Morejohn (1976). Evidence from seabirds of plastic particle pollution off Central California. *Western Birds*, 7 (3), 111–112.
Barnes, David K. A. (2002). Invasions by marine life on plastic debris. *Nature*, 416, 808–809.
Barnes, David K. A., Francois Galgani, Richard C. Thompson, and Morton Barlaz (2009). Accumulation and fragmentation of plastic debris in global

environments. *Philosophical Transactions of the Royal Society B: Biological Sciences*, 364 (1526), 1985–1998.

BIO Intelligence Service (2011). Assessment of impacts of options to reduce the use of single-use plastic carrier bags. Final report prepared for the European Commission – DG Environment. 19.04.2022 https://ec.europa.eu/environment/topics/waste-and-recycling/packaging-waste_en.

Brandão, Ana P., and Isabel Camisão (2021). Playing the market card: The Commission's strategy to shape EU cybersecurity policy. *Journal of Common Market Studies*, 1–21.

Carpenter, Edward J., and Kenneth L. Smith (1972). Plastics on the Sargasso Sea surface. *Science*, 175 (4027), 1240–1241.

Carpenter, Edward J., Susan J. Anderson, George R. Harvey, Helen P. Miklas, and Bradford B. Peck (1972). Polystyrene spherules in coastal waters. *Science*, 178 (4062), 749–750.

Clapp, Jennifer, and Linda Swanston (2009). Doing away with plastic shopping bags: International patterns of norm emergence and policy implementation. *Environmental Politics*, 18 (3), 315–332.

Colton, John B., Bruce R. Burns, and Frederick D. Knapp (1974). Plastic particles in surface waters of the Northwestern Atlantic: The abundance, distribution, source, and significance of various types of plastics are discussed. *Science*, 185 (4150), 491–497.

Copeland, Paul, and Scott James (2014). Policy windows, ambiguity and Commission entrepreneurship: Explaining the relaunch of the European Union's economic reform agenda. *Journal of European Public Policy*, 21 (1), 1–19.

Cornelius, Stephen E. (1975). Marine turtle mortalities along the Pacific coast of Costa Rica. *Copeia*, 1, 186–187.

Council of the European Union (2009). Note from General Secretariat to Delegations. 'Plastic Soup' – The problem of plastic marine debris in the oceans – Information from the Dutch Delegation. 14494/09. Brussels.

Cram, Laura (1994). The European Commission as a multi-organization: Social policy and IT policy in the EU. *Journal of European Public Policy*, 1 (2), 1–22.

Cundell, Anthony M. (1973). Plastic materials accumulating in Narragansett Bay. *Marine Pollution Bulletin*, 4 (12), 187–188.

Day, Robert H., Duff H. S. Wehle, and Felicia C. Coleman (1985). Ingestion of plastic pollutants by marine birds. In *Proceedings of the Workshop on the Fate and Impact of Marine Debris: 27–29 November 1984, Honolulu, Hawaii*, 344–386. 22.08.2021 https://repository.library.noaa.gov/view/noaa/5680.

Derraik, José G. B. (2002). The pollution of the marine environment by plastic debris: A review. *Marine Pollution Bulletin*, 44 (9), 842–852.

Dixon, Trevor R., and A. J. Cooke (1977). Discarded containers on a Kent beach. *Marine Pollution Bulletin*, 8 (5), 105–109.

Eckert, Sandra (2022). Circularity in the plastic age: Policymaking and industry action in the European Union. In Johanna Kramm and Carolin Völker (eds.). *Living in the plastic age. Perspectives from humanities, social sciences and environmental sciences.* Frankfurt, New York: Campus.

Eriksen, Marcus, Laurent C. M. Lebreton, Henry S. Carson, Martin Thiel, Charles J. Moore, Jose C. Borerro, Francois Galgani, Peter G. Ryan, and Julia Reisser (2014). Plastic pollution in the world's oceans: More than 5 trillion plastic pieces weighing over 250,000 tons afloat at sea. *PloS one*, 9 (12), 1–15.

European Commission (2010a). E-0825/10EN: Answer given by Mr Potočnik on behalf of the Commission. 11.02.2022 https://www.europarl.europa.eu/doceo/document/E-7-2010-0825-ASW_FR.html?redirect.

European Commission (2010b). Report of the workshop "Marine litter: Plastic soup and more", held on 8th November 2010 in Berlaymont, Brussels. 11.04.2022 https://ec.europa.eu/environment/marine/news-archive/index_en.htm#news Archive.

European Commission (2011a). Commission seeks views on reducing plastic bag use. IP/11/580. Brussels. 21.09.2021 https://ec.europa.eu/commission/presscorner/detail/en/IP_11_580.

European Commission (2011b). Communication from the Commission to the European Parliament, the Council, the European Economic and Social Committee and the Committee of the Regions. Roadmap to a resource efficient Europe. COM(2011) 571. Brussels. 22.09.2021 https://ec.europa.eu/environment/resource_efficiency/about/roadmap/index_en.htm.

European Commission (2012a). Commission staff working document. Overview of EU policies, legislation and initiatives related to marine litter. SWD(2012) 365. Brussels. 26.01.2022 https://ec.europa.eu/environment/marine/news-archive/index_en.htm#newsArchive.

European Commission (2012b). Environment: EU aiming to be at the forefront of efforts to reduce marine litter. Brussels. 26.01.2022 https://ec.europa.eu/commission/presscorner/detail/en/IP_12_1221.

European Commission (2013a). Environment: What should we do about plastic waste? New Green Paper opens EU-wide reflection. IP/13/201. Brussels. 21.09.2021 https://ec.europa.eu/commission/presscorner/detail/en/IP_13_201.

European Commission (2013b). Green Paper. On a European strategy on plastic waste in the environment. COM(2013) 123. Brussels. 03.05.2021 https://eur-lex.europa.eu/legal-content/EN/TXT/?uri=celex%3A52013DC0123.

European Commission (2014). Flash Eurobarometer 388. Attitudes of Europeans towards waste management and resource efficiency. Summary. 20.09.2021 https://data.europa.eu/data/datasets/s1102_388?locale=en.

European Commission (2015a). Communication from the Commission to the European Parliament, the Council, the European Economic and Social Committee and the Committee of the Regions. COM(2015) 614. Closing the loop – An EU action plan for the circular economy. Brussels. 04.05.2021

https://eur-lex.europa.eu/legal-content/EN/TXT/?uri=CELEX:52015DC06 14.
European Commission (2015b). Taking the EU resource efficiency agenda forward. A policymaker and business perspective. Luxembourg. 26.01.2022 https://ec.europa.eu/environment/enveco/resource_efficiency/.
European Commission (2017). Special Eurobarometer 468. Summary. Attitudes of European citizens towards the environment. 15.04.2022 https://europa.eu/eurobarometer/surveys/detail/2156.
European Commission (2018). Communication from the Commission to the European Parliament, the Council, the European Economic and Social Committee and the Committee of the Regions. A European strategy for plastics in a circular economy. COM(2018) 28 final. Brussels. 04.05.2021 https://eur-lex.europa.eu/legal-content/EN/TXT/?uri=CELEX:52018DC0028.
European Commission, and DG ENV (2011). In-Depth report. Plastic waste: Ecological and human health impacts. 20.09.2021 https://ec.europa.eu/environment/integration/research/newsalert/indepth_reports.htm.
European Commission, German Environment Agency and German Federal Ministry of the Environment, Nature Conservation and Nuclear Safety (2013a). Conference report. International conference on prevention and management of marine litter in european seas. 06.09.2021 https://www.muell-im-meer.de/aktivitaeten/international-conference-prevention-and-management-marine-litter-european-seas.
European Commission, German Environment Agency and German Federal Ministry of the Environment, Nature Conservation and Nuclear Safety (2013b). Message from Berlin. Conclusions of the chairpersons of the international conference on prevention and management of marine litter in European seas, held in Berlin, Germany, 10–12 April 2013. 06.09.2021 https://www.muell-im-meer.de/aktivitaeten/international-conference-prevention-and-management-marine-litter-european-seas.
European Parliament (2010). Recommendation on the proposal for a Council decision concerning the conclusion, on behalf of the European community, of the additional protocol to the cooperation agreement for the protection of the coasts and waters of the North-East Atlantic against pollution. A7-0009/2010. 28.09.2021 https://www.europarl.europa.eu/doceo/document/A-7-2010-0009_EN.html.
European Parliament (2013). Workshop. Plastic waste. PE 518.737. 20.09.2021 https://www.europarl.europa.eu/thinktank/en/document.html?reference=IPOL-ENVI_AT%282013%29518737.
European Parliament (2014). Plastic waste in the environment. European Parliament Resolution of 14 January 2014 on a European strategy on plastic waste in the environment. 03.05.2021 https://op.europa.eu/en/publication-detail/-/publication/e1c303f7-c8a2-11e6-a6db-01aa75ed71a1/.

European Parliament, and Council of the European Union (2015). Directive (EU) 2015/720 of the European Parliament and of the Council of 29 April 2015 Amending Directive 94/62/EC as regards reducing the consumption of lightweight plastic carrier bags. 03.05.2021 https://eur-lex.europa.eu/legal-content/en/TXT/?uri=celex%3A32015L0720.

Farmer, Andrew M. (2012). Overview of EU policy: Waste. In Andrew M. Farmer (ed.) *Manual of European Environmental Policy*, London: Routledge. 10.09.2021 https://ieep.eu/understanding-the-eu/manual-of-european-environmental-policy.

Frisch Aviram, Neomi, Nissim Cohen, and Itai Beeri (2020). Wind(ow) of change: A systematic review of policy entrepreneurship characteristics and strategies. *Policy Studies Journal*, 48 (3), 612–44.

Glaser, Barney G., and Anselm L. Strauss (2017). *The Discovery of Grounded Theory: Strategies for Qualitative Research*. New York: Routledge.

Gold, Mark, Katie Mika, Cara Horowitz, Megan Herzog, and Lara Leitner (2013). Stemming the tide of plastic marine litter. A global action agenda. Pritzker Brief No. 5. 01.09.2021 https://escholarship.org/uc/item/6j74k1j3.

Gregory, Murray R. (1978). Accumulation and distribution of virgin plastic granules on New Zealand beaches. *New Zealand Journal of Marine and Freshwater Research*, 12 (4), 399–414.

Gregory, Murray R. (2009). Environmental implications of plastic debris in marine settings – Entanglement, ingestion, smothering, hangers-on, hitch-hiking and alien invasions. *Philosophical Transactions of the Royal Society B: Biological Sciences*, 364 (1526), 2013–2025.

Happaerts, Sander (2014). International discourses and practices of sustainable materials management. Leuven. 20.09.2021 https://ce-center.vlaanderen-circulair.be/en/publications.

Harper, P.C., and J. A. Fowler (1987). Plastic pellets in New Zealand storm-killed prions *(Pachyptila* spp.) 1958-1977. *Notornis*, 34, 65–70. https://www.birdsnz.org.nz/publications/plastic-pellets-in-new-zealand-storm-killed-prions-pachyptila-spp-1958-1977-2/.

Herweg, Nicole (2013). Der Multiple-Streams-Ansatz – Ein Ansatz, dessen Zeit gekommen ist? *Zeitschrift für Vergleichende Politikwissenschaft*, 7, 321–345.

Herweg, Nicole (2016). Explaining European agenda-setting using the Multiple Streams Framework: The case of European natural gas regulation. *Policy Sciences*, 49, 13–33.

Hidalgo-Ruz, Valeria, and Martin Thiel (2015). The contribution of citizen scientists to the monitoring of marine litter. In Michael Klages, Lars Gutow and Melanie Bergmann (eds.). *Marine anthropogenic litter*, 429–447. Cham/Heidelberg/New York/Dordrecht/London: Springer.

Iusmen, Ingi (2013). Policy entrepreneurship and eastern enlargement: The case of EU children's rights policy. *Comparative European Politics*, 11 (4), 511–529.

Jones, Bryan D. (1994). *Reconceiving decision-making in democratic politics: Attention, choice, and public policy.* Chicago: University of Chicago Press.

Jones, Michael D., Holly L. Peterson, Jonathan J. Pierce, Nicole Herweg, Amiel Bernal, Holly Lamberta Raney, and Nikolaos Zahariadis (2016). A river runs through it: A multiple streams meta-review. *Policy Studies Journal*, 44 (1), 13–36.

Kenyon, Karl W., and Eugene Kridler (1969). Laysan albatrosses swallow indigestible matter. *The Auk*, 86 (2), 339–43.

Kingdon, John W. (2014). *Agendas, alternatives, and public policies.* Harlow: Pearson.

Knaggård, Åsa (2015). Forum section: Theoretically refining the Multiple Streams Framework. The Multiple Streams Framework and the problem broker. *European Journal of Political Research*, 54 (3), 450–465.

Kramm, Johanna, and Carolin Völker (2018). Understanding the risks of microplastics: A social-ecological risk perspective. In Martin Wagner and Scott Lambert (eds.). *Freshwater microplastics: Emerging environmental contaminants?*, 223–237. Cham: Springer.

Langer, Magdalena (2017). Plastic discourses: A discursive agency perspective on plastic waste policy in the European Union. Master Thesis, Albert-Ludwig-University Freiburg (unpublished).

Leipold, Sina (2021). Transforming ecological modernization 'from within' or perpetuating it? The circular economy as EU environmental policy narrative. *Environmental Politics*, 30 (6), 1045–1067.

Maltby, Tomas (2013). European Union energy policy integration: A case of European Commission policy entrepreneurship and increasing supranationalism. *Energy Policy*, 55 (100), 435–444.

Mederake, Linda, and Doris Knoblauch (2019). Shaping EU plastic policies: The role of public health vs. environmental arguments. *International Journal of Environmental Research and Public Health*, 16 (20), 1–18.

Menz, Georg (2013). The European Commission as a policy entrepreneur in european migration policy making. *Regions & Cohesion*, 3 (3), 86–102.

Moore, Charles J. (no date). History: "How a chance encounter with plastic in the North Pacific Ocean changed my life's work". 15.09.2021 https://www.captain-charles-moore.org/about.

Moore, Charles J., Susan L. Moore, Molly K. Leecaster, and Stephen B. Weisberg (2001). A comparison of plastic and plankton in the North Pacific central gyre. *Marine Pollution Bulletin*, 42 (12), 1297–1300.

Parslow, John L. F., and Don Jefferies (1972). Elastic thread pollution of puffins. *Marine Pollution Bulletin*, 3 (3), 43–45.

PlasticsEurope (2021). Plastics – the Facts 2021. An analysis of European plastics production, demand and waste data. 23.04.2022 https://plasticseurope.org/knowledge-hub/plastics-the-facts-2021/.

Potočnik, Janez (2010). Marine litter. From shelf to shore. SPEECH/10/626. Brussels. 06.09.2021 https://ec.europa.eu/commission/presscorner/detail/en/SPEECH_10_626.

Potočnik, Janez (2011). Let's keep the Mediterranean litter-free. SPEECH/11/247. Athens. 21.09.2021 https://ec.europa.eu/commission/presscorner/detail/en/SPEECH_11_247.

Potočnik, Janez (2012). Any future for the plastic industry in Europe? SPEECH/12/632. Wiesbaden. 21.09.2021 https://ec.europa.eu/commission/presscorner/detail/en/SPEECH_12_632.

Potočnik, Janez (2013). Let's free our oceans from this plague. SPEECH/13/312. Berlin. 06.09.2021 https://ec.europa.eu/commission/presscorner/detail/en/SPEECH_13_312.

Princen, Sebastiaan (2009). *Agenda-Setting in the European Union.* London: Palgrave Macmillan UK.

Rothstein, Stephen I. (1973). Plastic particle pollution of the surface of the Atlantic Ocean: Evidence from a seabird. *The Condor*, 75 (3), 344–345.

Ryan, Peter G. (2015). A brief history of marine litter research. In Michael Klages, Lars Gutow and Melanie Bergmann (eds.). *Marine anthropogenic litter*, 1–28. Cham/Heidelberg/New York/Dordrecht/London: Springer.

Ryan, Peter G., Charles J. Moore, Jan A. van Franeker, and Coleen L. Moloney (2009). Monitoring the Abundance of plastic debris in the marine environment. *Philosophical Transactions of the Royal Society B: Biological Sciences*, 364 (1526), 1999–2012.

Schmidt, Susanne K. (2001). Die Einflussmöglichkeiten der Europäischen Kommission auf die Europäische Politik. *Politische Vierteljahresschrift*, 42 (2), 173–192.

Schön-Quinlivan, Emmanuelle, and Marco Scipioni (2017). The Commission as policy entrepreneur in European economic governance: A comparative multiple stream analysis of the 2005 and 2011 reform of the stability and growth pact. *Journal of European Public Policy*, 24 (8), 1172–1190.

Simon, Nils, Doris Knoblauch, Linda Mederake, Katriona McGlade, Maro L. Schulte, and Supriya Masali (2018). No more plastics in the ocean. Gaps in global plastic governance and options for a legally binding agreement to eliminate marine plastic pollution. 25.08.2021 https://www.adelphi.de/en/publication/no-more-plastics-ocean.

Simon, Nils, and Maro L. Schulte (2017). Stopping global plastic pollution. The case for an international convention. Berlin. 20.09.2021 https://www.adelphi.de/en/publication/stopping-global-plastic-pollution-case-international-convention.

Thompson, Richard C., Ylva Olsen, Richard P. Mitchell, Anthony Davis, Steven J. Rowland, Anthony W. G. John, Daniel McGonigle, and Andrea E. Russell (2004). Lost at sea: Where is all the plastic? *Science*, 304 (5672), 838–838.

United Nations Environment Programme (2005). Marine litter. An analytical overview. 25.08.2021 https://wedocs.unep.org/handle/20.500.11822/8348.

Winston, Judith E. (1982). Drift plastic – An expanding niche for a marine invertebrate? *Marine Pollution Bulletin*, 13 (10), 348–351.

Wong, Charles S., David R. Green, and Walter J. Cretney (1974). Quantitative tar and plastic waste distributions in the Pacific Ocean. *Nature*, 247, 30–32.

Zahariadis, Nikolaos (2007). The Multiple Streams Framework: Structure, limitations, prospects. In Paul A. Sabatier (ed.). *Theories of the policy process*, 65–92. Cambridge: Westview Press.

Zahariadis, Nikolaos (2008). Ambiguity and choice in European public policy. *Journal of European Public Policy*, 15 (4), 514–530.

Zahariadis, Nikolaos (2016). Setting the agenda on agenda setting: Definitions, concepts, and controversies. In Nikolaos Zahariadis (ed.). *Handbook of public policy agenda setting*, 1–22. Cheltenham, UK/Northampton, MA, USA: Edward Elgar Publishing.

Zahariadis, Nikolaos, and Theofanis Exadaktylos (2016). Policies that succeed and programs that fail: ambiguity, conflict, and crisis in Greek higher education. *Policy Studies Journal*, 44 (1), 59–82.

Microplastic in the Aquatic Environment

Maren Heß, Carolin Völker, Nicole Brennholt, Pia Maria Herrling, Henner Hollert, Natalia P. Ivleva, Jutta Kerpen, Christian Laforsch, Martin Löder, Sabrina Schiwy, Markus Schmitz, Stephan Wagner, Thorsten Hüffer

Introduction

While the increasing load of plastic waste in the environment has been observable for many decades, the smaller and less obvious microplastics have only come into focus in recent years. Although marine studies from the 1970s (e.g., Carpenter et al. 1972; Colton et al. 1974) reported an increasing input of small plastic particles into the oceans, systematic research on this topic started only about three decades later. In one of the initializing studies, Thompson et al. (2004) asked, "Lost at Sea: Where is all the Plastic?" and brought the fragmentation of plastic waste into a continuum of macro-, micro-, and nanoplastics into the scientific, political, and public focus—the latter at least by introducing the buzzword "microplastic(s)". From that point, the research on microplastics has broadened from marine to freshwater and later also to terrestrial ecosystems as well as the atmosphere. Microplastics have been detected in every environmental compartment and almost all ecosystems around the world.

Research is already underway on developing potential substitutes, such as biodegradable plastics, and regulators debate measures including bans on certain products or the restriction of microplastics in products placed on the market (ECHA 2019). At the same time, there is still a lack of scientific knowledge about the fate of microplastics in the environment and its long-term consequences. A consensus on a science-based definition and categorization of environmental plastic debris (Hartmann et al. 2019; Rochmann et al. 2019), the harmonization of analytical tools to assess environmental concentrations, and quality control and assurance protocols for experiments studying the environmental fate, transport, and effects of micro- and nanoplastics are the first necessary steps in this regard. Generally, accepted frameworks are essential prerequisites for a comprehensive environmental risk assessment and for a dialogue on regulatory measures.

Scientific knowledge about entry pathways, distribution, and accumulation in the environment, as well as the effects of microplastics on organisms and ecosystems is the basis for precise and legitimate measures. A profound scientific knowledge base is therefore the foundation for all further discussions on this topic.

This chapter aims to provide an overview of the current state of knowledge on micro- and nanoplastics from an environmental science perspective. The most important research areas and findings are presented, but also open knowledge gaps and challenges are identified, e.g., in the development of sufficiently sensitive analytical or bioassay methods. Finally, the aim is to illustrate the complexity of the topic and why hypothesis-driven research is necessary to address it—but, however, is time intensive.

Towards a Definition for Microplastics

The term "microplastics" has been a catchphrase for various kinds of small plastic particles. Microplastics are often thought of as one class of contaminants with distinct properties such as many dissolved cchemicals (e.g., polycyclic aromatic hydrocarbons, polychlorinated biphenyls, and many more). However, "microplastics" cover a broad spectrum of intrinsic chemical and physical properties including polymer types, sizes, shapes, and the presence of plastic additives (Fig. 1). This makes microplastics rather a collective term covering diverse particles with diverse molecular features from a broad range of applications (Rochmann et al. 2019). There is no consensus so far on which of these various properties define a particle as microplastic. There is an internationally valid definition for plastic as a

"material which contains as an essential ingredient a high molecular weight polymer and which, at some stage in its processing into finished products, can be shaped by flow" (ISO 2013).

However, this definition of plastics is purely based on material science and excludes materials such as some elastomers (e.g., rubber). Tire materials, which consist mainly of rubber, have long been left out of the discussion and research on microplastics in the environment due to different material properties and the original extraction from natural sources (rubber tree). In fact, tire wear particles represent a substantial fraction of synthetic polymer-based particles in the environment (Lassen et al. 2015). Hence, tire wear particles have been proposed to be included in the definition of environmental plastics (Hartmann et al. 2019). An agreement on an internationally valid definition would be an indispensable prerequisite for conducting comparable scientific investigations on identical particle types.

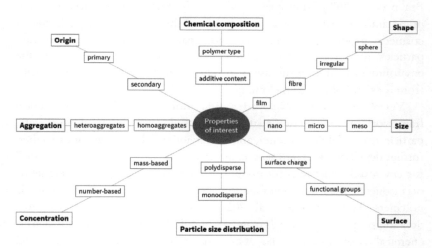

Fig. 1: *Spectrum of intrinsic chemical and physical properties of plastic particles in the environment. (Source: adopted from Hasselöv and Kaegi 2009)*

Categorization according to the varying properties of the particles would be appropriate to reflect the complexity of the properties of microplastics and their different environmental fates (e.g., buoyancy/sedimentation, degradation/persistence, and uptake by organisms). Due to this diversity, scientific investigations can often only cover part of the plastic particles or their effects (e.g., sampling buoyant particles at the water surface or sinking particles in the water column/sediment or effects of spheric vs. fibrous particles). It is essential to name and differentiate specific categories, thereby enabling better comparability of research results from different studies and a more targeted derivation of measures.

The most prominent criterion and the first one to be introduced in the scientific literature by NOAA in 2008 was particle size. The suggested upper size limit of 5 mm is widely accepted (Arthur et al. 2008). No lower size limit or further classification characteristics beyond size were established in early microplastic studies. It should be noted that there are now proposals not only to categorize different size classes but also to subdivide nanoplastics. Currently, plastic particles and fibers smaller than 1 μm and in the size range 1 μm to 1 mm are defined as nanoplastics and microplastics, respectively (Hartmann et al. 2019; ISO/TR 21960; Braun et al. 2020). Fragments in the size range 1–5 mm can be referred to as large microplastics (ISO/TR 21960; Braun et al. 2020). Microplastics were early differentiated based on their origin into primary particles produced in this size class (e.g., beads in cosmetic products, granules for industrial processing) and secondary particles produced by fragmentation of larger plastic products in the environment or during the use of products, such as the abrasion particles from tires and fibers from clothing.

Verschoor (2015) developed a first more comprehensive definition framework using five criteria such as chemical composition, physical state, particle size, solubility in water, and degradability. This framework was then further developed by Hartmann et al. (2019) using the term "plastic debris" for environmentally relevant plastics to emphasize that these plastic items occur outside their intended function and to be distinct from water-soluble polymers (Huppertsberg et al. 2020). The framework consists of three defining properties by which materials should be considered plastic debris: chemical composition, solid state, and solubility. The five additional categories consisting of particle size, shape and structure, color, and origin can be used for further classification. Although applying eight criteria sounds logical, simple, and rather straightforward, defining the boundaries is highly complex and requires further discussion and the search for consensus.

One of the key challenges regarding the chemical composition lies within the origin of the polymer. Typically, one would distinguish between a polymer of natural origin (e.g., DNA, cellulose, wool) or synthetic (e.g., nylon, polyethylene, polyester, Teflon, and epoxy). One difficulty is to delineate at what point natural polymers can be classified as artificially modified to the extent that they can or must be included in the definition of plastic debris. Examples are highly modified natural rubber or cellulose to rayon/viscose and cellophane. While the dyeing of wool represents an artificial treatment, it does not change the core of the material, the natural polymer—and would therefore not be assigned to plastic debris. Another challenge poses the classification of composite materials or plastic materials that contain a high proportion of additives. The additives give the materials their final properties, such as flexibility (plasticizers), stability (stabilizers), or fire resistance (flame retardants). Depending on the plastic, the proportions can even be >50 percent and thus the material consists largely of low-molecular weight chemicals and not polymers. The discussion on the criteria for a definition of environmental plastic debris should be intensified and broadened beyond the scientific community (e.g., for regulatory purposes) (Brennholt et al. 2018). The International Organization for Standardization (ISO) has established an ad hoc working group "Microplastics" working on a valid definition and nomenclature as a precondition for further standardization of monitoring and test methods (for example: ISO/TR 21960).

Analysis of Microplastics in the Aquatic Environment

Although harmonization and standardization of the detection of microplastics in environmental samples has been sought for some time at both national and international levels, comparability between individual studies is still not possible due to methodological differences (Hidalgo-Ruz et al. 2012; Löder and Gerdts 2015). The scientific community now agrees that microplastic classification based on purely optical criteria is insufficient because of the occurrence of very high error rates, especially for smaller particles. Reliable detection of microplastics can only be achieved by chemical characterization (Ivleva 2021; Löder and Gerdts 2015). Although a few other analytical approaches exist, there are two major methodological pathways for chemical characterisation of microplastics—particle-related analytical techniques and mass-related analytical techniques—on which we will focus in this section. Both ways for analysis of microplastics should be seen complementary and their application depends on the respective research question. The characterization of plastic particles is mainly important for ecology or ecotoxicology, while mass balances are more interesting for water management. Particle-related techniques like infrared spectroscopy, e.g., Fourier transform infrared spectroscopy (FTIR) (Löder et al. 2015; Schymanski et al. 2021), and Raman spectroscopy (Raman) (Anger et al. 2018; Schymanski et al. 2021), can be used for ecological and ecotoxicological research questions because size, shape and number of particles matter. Mass-related techniques like pyrolysis gas chromatography/mass spectroscopy (Pyr-GC/MS) (Fischer and Scholz-Böttcher 2017) or thermal extraction/desorption gas chromatography/mass spectroscopy (TED-GC/MS) (Dümichen et al. 2017) enable to establish mass balances, which is important for water management and specific modeling approaches

Contamination of samples with microplastic particles from ambient air, clothing, chemicals or laboratory utensils used can be a significant problem in microplastic analysis of environmental samples. For this reason, cotton lab coats should be worn and, wherever possible, plastic laboratory utensils (including bottle lids) should be replaced with glass, metal, or rare plastics such as polytetrafluoroethylene. Laboratory work should take place under a laminar flow box and vessels containing samples should be covered with glass lids or aluminum foil at all times. To avoid contamination, all required reagents and liquids should further be filtered before use and all laboratory equipment should thoroughly be rinsed with pre-filtered deionized water, 35 percent ethanol and again with water before use and between steps. Blank samples should be processed in parallel with the environmental samples to monitor for possible contamination and to correct the data.

Sampling of Microplastics

Microplastics are particulate contaminants that are mostly not homogeneously distributed in aquatic environments. Next to polymer type, density, shape, size, and chemical additives, microplastics in the environment are subject to aging, biofouling, and aggregation (Fig. 1). All these particle-related factors in tandem with physical factors including wind, currents, turbulence, and mixing present in the respective environment, affect the distribution of microplastics in aquatic habitats and complicate sampling in aquatic environments.

Microplastics are now detected in all waters, but due to relatively low concentrations in the samples, it is necessary to concentrate the particles during or following sampling. Plankton nets are mostly used and deployed as so-called "manta nets" at the water surface to collect floating microplastics and efficiently filter water samples in the range of several thousand liters. This allows for good capture of the larger, but less abundant microplastics. Due to the typical mesh size of 300 μm[1], only particles larger than the mesh size can be quantified. The smaller, ecologically more questionable microplastics pass through the mesh. For this reason, a two-pronged approach to water sampling has now been established using a large-area filtration unit with a 500 μm prefilter, which concentrates the more common microplastics <500 μm on stainless steel sieve mesh, with a mesh size of 10 μm, filtering several hundred liters. Thus, the broad microplastic size range of 10 μm–5 mm is completely quantified. Alternatives for the detection of smaller microplastics are the use of candle filter units or sieve cascades for the filtration of water samples (Mintenig et al. 2017, 2019).

Liedermann et al. (2018) pointed out that plastic transport is not limited to the surface layer of a river, and must be examined within the whole water column as well as for suspended sediments. The implementation of repeated one-location multi-point microplastic surveys for sampling simultaneously at the water surface, in the water column, the water/sediment interface, and in the sediment promise a mechanistic understanding of spatial microplastic distribution. Such a survey design also confronts researchers with a number of samples that is hard to manage, at least when targeting a single-particle resolution.

Sample Purification (for Spectroscopic Methods)

The analysis of small microplastics (<500 μm) with spectroscopic methods like micro-FTIR or Raman spectroscopy necessarily requires the placement of samples on measurement substrates such as filter membranes (Käppler et al. 2015). An unfavorable target (microplastics)-non-target (organic or inorganic particles) ratio mostly prevails in water samples, which often makes both filtration and spectroscopic analysis simply impossible (Löder et al. 2017). The purification of microplastic samples is an essential part of sample preparation and a crucial step before the actual analysis; and it is also of advantage for the available GC/MS techniques (Fischer and Scholz-Böttcher 2017).

As a prerequisite, the ideal purification approach should efficiently remove the organic and inorganic materials from the samples, allow the concentration of the purified samples on filters with a small pore size (<1

---

1 1 micrometer = 10^{-6} meter = 1000 nanometer

µm), be inexpensive and low effort, and at the same time not interfere with the different synthetic polymers. To date, a wide variety of purification methods for microplastic samples have been developed, often using strongly acidic or alkaline solutions, oxidizing agents, or combined approaches. Especially more sensitive synthetic polymers can be lost or fragmented during digestion with the more aggressive approaches (Löder et al. 2017).

Alternatively, a plastic-friendly and effective approach for the purification of environmental samples is using a combined enzymatic-oxidative approach to obtain the complete spectrum of available plastics in the samples (Löder et al. 2017).

Identification and Quantification of Microplastics

The diversity and complexity of plastic sources, usage patterns, emission pathways, and material properties is reflected in the diversity of microplastics, exhibiting a high variety of physical, chemical, and biological characteristics (e.g., size, shape, density, polymer type, surface properties, etc.). Therefore, advanced methods are required for the reliable identification and quantification of this analyte[2], probably one of the most challenging analytes in the aquatic environment (Ivleva 2021). An overview of applicable methods is given in Figure 2.

---

2 The term analyte refers to a sample of a material that is being analyzed.

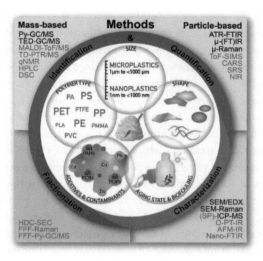

Fig. 2: Complexity and diversity of micro- and nanoplastics and methods applied for the analysis. (Source: Ivleva 2021)

Particle-Related Analytical Techniques

Net samples can be concentrated in the laboratory on a 500 µm stainless steel sieve. Any potential microplastic particle in the sieve residue is then sorted out visually under a stereomicroscope, photo-documented and measured. This size class can be easily handled with tweezers and plankton counting chambers, e.g. Bogorov chambers, facilitate systematic sorting. The individual plastic particles can be identified using various spectroscopic methods, e.g., attenuated total reflection (ATR) FTIR spectroscopy or Raman spectroscopy (Anger et al. 2018; Schymanski et al. 2021), to highlight only the most common methods. Both spectroscopic techniques allow for matching the spectra of the particles with references in less than a minute.

The purified particles <500 µm from filtration set-ups are deposited on e.g. alumina filters for FTIR or on Au-coated PC filters for Raman measurement (Schymanski et al. 2021). When using FTIR, the complete sample filter is measured by e.g., focal plane array (FPA) detector, for chemical imaging under the IR microscope in transmission mode (Löder et al. 2015). To cover the complete sample, it is scanned—FPA area by FPA tile—one after the other. The resulting chemical imaging data comprise million spectra and more, making manual evaluation based on characteristic plastic bands very time-consuming. For this reason, automated analysis is essential. Here, recently developed free-of-charge software tool, allowing the systematic identification of MPs in the environment (siMPle) can be applied (Primpke et al. 2020). Alternatively, novel software solutions based on Random Decision Forest Classifiers (RDFC) which allow automated analysis of such data in a few minutes (Hufnagl et al. 2019, 2022) can be used. For final quality assurance, the obtained list of results is only checked manually—an immense time saving compared to purely manual analysis.

Raman spectroscopy (similar to IR spectroscopy) provides vibrational fingerprint spectra, and allows for identification of plastic particles and some additives (e.g., pigments, oxides) as well as other (in)organic and (micro)biological compounds. The combination of Raman spectroscopy with confocal optical microscopy (Raman microspectroscopy or μ-Raman spectroscopy) and the application of excitation lasers in the visible range enables a significantly better spatial resolution, i.e., down to 1 μm and even below (down to approximately 300 nm) compared to μ-(FT)IR spectroscopy (where spatial resolution of approximately 10 μm can be achieved). Therefore, μ-Raman spectroscopy can be recommended, especially for the analysis of plastic particles smaller than 10−20 μm (Ivleva 2021).

In contrast to IR, the Raman-based methods offer the advantage of insensitivity toward water enabling investigations of microplastics in aqueous and (micro)- biological samples (2D and 3D chemical imaging). However, the μ-Raman analysis is usually more time-consuming and often suffers from the fluorescence interference (which can be minimized by the proper sample preparation and optimization of measurement parameters). To analyze representative numbers of particles on the (entire) filter, the automated particle recognition approach, utilizing optical image with the subsequent μ-Raman measurements in particle-by-particle mode has been developed (TUM-ParticleTyper (von der Esch et al. 2020) and GEPARD (Brandt et al. 2020)). IR and Raman spectroscopy, being complementary, can be efficiently applied for detailed analysis of microplastic samples.

Mass-Related Analytical Techniques

Pyr-GC/MS or TED-GC/MS can be efficiently applied to obtain information about the chemical composition of potential microplastic particles in environmental samples by analysing their thermal degradation products. Furthermore, plastic additives can be determined simultaneously during Pyr-GC/MS analysis if a thermal desorption step precedes the actual pyrolysis (Fries et al. 2013). Original Pyr-GC/MS approaches involved manual handling and measurement of single particles and as only one particle per run could be analysed, the technique was thus not suitable for processing large amounts of samples. Latest Pyr-GC/MS developments facilitated the measurement of whole environmental samples concentrated on filters which relativize the above mentioned restrictions (Fischer and Scholz-Böttcher 2017). Similar GC-MS technique, namely TED-GC/MS, allows for the analysis of a significantly larger sample amount (mg-range compared to µg-range by Pyr-GC/MS). In TED-GC/MS the degradation products of synthetic polymers are adsorbed on a solid-phase adsorber and subsequently transferred to the analysis platform (Dümichen et al. 2017). The solid-phase adsorber is loaded after organic material is decomposed, thus the microplastic analysis in highly polluted samples (containing more than 0.5–1 percent by weight for each type of polymer analyzed) can be performed without preceding purification of the sample. As the expected microplastic content in the most real environmental samples is below one percent by weight, microplastics preconcentration per (cascade) filtration and removal of accompanying (in)organic matrices will improve the detection and quantification sensitivity of both Py-GC/MS and TED-GC/MS (as well as spectroscopic methods, as discussed above) for the analysis of environmental samples (Braun et al. 2020). An overview of methodical parameters for particle- and mass-related methods for MP analysis is given in Table 1 (Braun et al. 2020).

A combination of both analytical methods would be ideal, but is time-consuming and costly. Currently, spectroscopic methods for quantifying and characterizing particles are used rather for questions on ecological and ecotoxicological effects, and mass balance methods for water management questions (e.g. performance of wastewater treatment plants, total load in streams or in the sea).

Tab. 1: *Overview of methodical parameters for detection methods, without consideration of the sample composition in particular (organic/inorganic) and sample preparation; abbreviation of methods. (Source: Braun et al. 2020)*

Properties	Particle-related techniques				Mass-related techniques	
	µ-Raman	IR / FTIR (µ-/FPA-)	µ ATR-FTIR	ATR-FTIR	Py-GC/MS*	TED-GC/MS
Sample template for measurement	Prepared filter residue	Prepared filter residue	Isolated particles	Isolated particles	Isolated particles / prepared sample	Sample
Maximum particle number per filter/mass in sample	$10^3 - 10^5$	$10^3 - 10^5$	undefined	undefined	µg	mg
Dimension measuring time (real environ. sample)	d - h	d - h	min	min	h	h
Lower detection limit (in practice)	1 - 5 µm	10 µm	25 - 50 µm	500 µm	0.01 - 1 µg (abs)**	0.5 - 2.4 µg (abs.)**

* *Various superstructures of the pyrolysis unit (e.g. Curie point, filament, micro furnace).*
** *depending on polymer type and pyrolysis unit.*

Transport, Distribution, and Retention of Microplastics in Aquatic Systems

Emissions and Emission Pathways

In Germany, the highest emissions of microplastics are attributed to traffic, infrastructure, and buildings (62 percent), followed by the private and industrial sectors (24 percent) (Bertling et al. 2018). The share of different sources on total plastic emissions is highly dependent on the geographical location (Meijer et al. 2021). Hence, no globally uniform emission patterns can be derived for plastics.

Depending on their origin, microplastics detected in the environment can be shaped as spheres, fragments, and fibers (Fig. 1). Fibers are commonly released from textiles e.g., during laundry. Fragments are generated during abrasion and fragmentation of larger plastic items. Spherical microplastics, often referred to as pellets or beads, are intentionally produced particles with a defined particle size and shape (Hartmann et al. 2019). Plastic emissions may occur via point sources and diffusive sources in terrestrial and aquatic environments. Wastewater discharge, in particular combined sewer overflow, is the predominant point source for aquatic environments (Dris et al. 2018). Diffuse emissions include atmospheric transport, but also flooding and littering as well as road runoff may occur.

The highest share of total plastic emissions is attributed to tire wear with emission estimates between 75×10^3 to 130×10^3 tons for Germany and 500×10^3 to 1.300×10^3 tons for the EU. First studies indicate major tire road wear particle sinks are roadside soils (36–76 percent) and freshwater sediments (10–22 percent) (Bänsch-Baltruschat et al. 2020). Major sources of microplastics pollution in agricultural soils include application of biosolids and compost, wastewater irrigation, mulching film, polymer-based fertilizers, and pesticides, and atmospheric deposition. For Switzerland, it was estimated that macro- and microplastic emissions to soil are up to 40-times higher than to the aquatic environment (Kawecki and Nowack 2019). It is estimated that 32 percent of plastics produced are in the terrestrial and not in the aquatic environment (Kumar et al. 2020). Marine plastic pollution is dominated by land-based plastics, either by direct emission from coastal zones or by transport through rivers (Fig. 3).

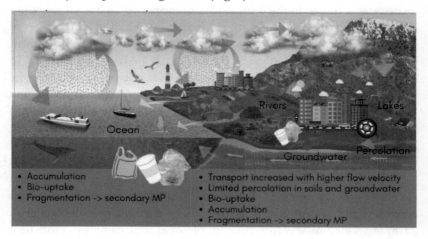

Fig. 3: Microplastics in the water cycle. (Source: own representation)

Recent studies quantified plastic pollution in rivers and related emissions and transport to human settlements, to sewer systems and to flood events (Roebroek et al. 2021; Meijer et al. 2021). Observed plastic concentrations and loads in rivers depend not only on the magnitude of emission but also on instream processing including storage, remobilization, sorting, and fragmentation. A better understanding of pathways and transport mechanisms of plastic waste to and within rivers and the global distribution of riverine plastic emissions into the ocean will contribute to develop effective prevention and collection strategies.

Transport and Distribution in Aquatic Systems

Wastewater effluents, combined sewer overflows, stormwater drain outlets, and littering contribute to plastic pollution of rivers (Wagner et al. 2019), (Fig. 4). For instance, ten-year return period floods already tenfold the global plastic mobilization potential compared to non-flood conditions (Roebroek et al. 2021). Rainwater runoff and wastewater is transported in the sewer system to the wastewater treatment plant or directly discharged into surface water bodies. Regarding urban sewage systems, there are two types: 1) a separate sewer system for rainwater and sewage water and 2) a combined sewage system where sewage water and stormwater are conveyed in the same pipe to the treatment plant for treatment. In the separate system, only the wastewater is treated in the wastewater treatment plant, while the rainwater runoff is discharged directly to the water bodies. The percentage of wastewater and stormwater, which is not treated in wastewater treatment plants, is estimated by Bertling et al. (2018) at 22 percent for Germany. In both systems, there are also numerous sinks for microplastics beside the wastewater treatment plants, such as street inlets on traffic areas, infiltration systems, rainwater retention and overflow basins, and soil retention filters. Data about microplastic pollution of stormwater are rarely available, but high microplastic loads may occur during heavy rainfall events as indicated by some recent studies.

In municipal or industrial wastewater treatment plants, microplastics are eliminated by aggregation with sewage sludge flocs with a treatment efficiency >95 percent related to the mass of microplastics (Bertling et al. 2018). An additional downstream sand filtration stage eliminates more than 99 percent of microplastics (Wolff et al. 2021). If the sewage sludge is incinerated, microplastics are oxidized to carbon dioxide. In cases where the sewage sludge is used as fertilizer in agriculture or put to landfill, microplastics are released into the soil and could consequently get re-mobilized into the water cycle.

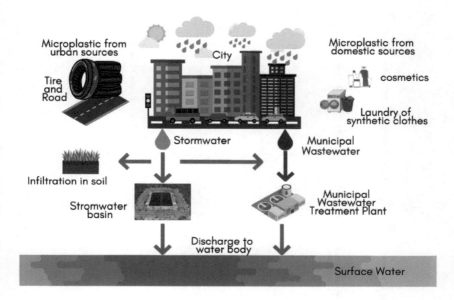

Fig. 4: Emission pathways of urban microplastic in surface waters. (Source: own representation)

Concentration and Distribution of Microplastics in Surface Waters

The concentration and distribution of microplastics on the water surface, in the water column and sediment of surface waters showed that their occurrence depends on many variables: geographical position, wind, currents, plastic properties like density and size but also on stream flow rate (Bellasi et al. 2020). Detected microplastic particle numbers do not merely depend on sampling techniques, sample preparation procedures, and analytical methods, but also on sampling location (i.e., urban vs. rural areas, near wastewater treatment plants, size and characterization of catchment area, water surface vs. water column, sediment depth), season of the year, and environmental conditions during sampling (i.e., heavy rainfall or flooding events). Microplastic can accumulate in certain compartments of the water bodies.

Exemplarily, 41 environmental studies published between 2011 and 2019 were assessed for comparing aquatic environmental concentrations in the

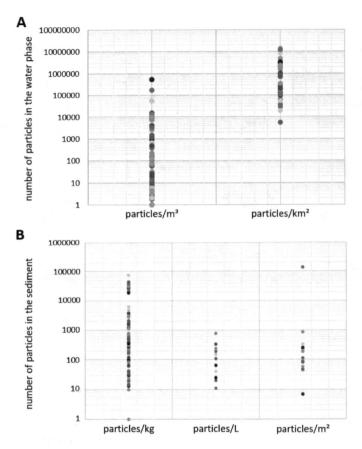

Fig. 5: Environmental concentrations of microplastic particles in (A) the water phase of rivers and lakes around the world based on 41 studies published between 2011 and 2019 and (B) river and lake sediments around the world based on 25 studies published between 2014 and 2019 chosen as an example. Each study investigated a variable number of sampling sites. Each point represents one sampling site. (Source: own draft, similar to Scherer et al. 2020)

water phase of rivers and lakes (similar to Scherer et al. 2020, Fig. 5). Most of these studies presented their results in particles per volume ("particles per m^3": n=26, "particles per L": n=9) and some studies showed their results relating to the water surface ("particles per km^2": n=5). Just one study used a weight indication ("g per 1,000 m^3"). Here the data given vary between

0.029 and 0.516 g/1,000 m^3. For more clarity in Figure 5, the unit "particles per L" has been converted into "particles per m^3" and the only study using a weight indication is not shown. The measured particle numbers show a wide variability and range from one to about one million particles per m^3.

The situation is similar for studies in sediments that use particles/kg, particles/L, or particles/m^2—making them incomparable. Here the data vary between one and about 100,000 particles per kg of sediment. In summary therefore, not only different methods applied making a direct comparison between these studies difficult, but also the unit in which quantities are shown.

Weathering and Degradation

Once plastics are emitted to the environment, they are immediately subjected to environmental factors including light, humidity, temperature, mechanical stress and biota causing alterations of the plastic surface, changes in the chemical bulk properties like composition but also physical properties like particle size (Arp et al. 2021). These processes are referred to as weathering and proceed along two interconnected and often synergistic routes: first, fragmentation and the release of soluble or volatile components, and second, biofouling and oxidative degradation (MacLeod et al. 2021). Photooxidation of plastics alters the surface properties like hydrophobicity, the composition due to the consumption of antioxidants, and consequently leads to embrittlement of the material. This process is a prerequisite for fragmentation of plastics where small particles are released from larger plastic items forming secondary microplastics (Song et al. 2017). Composition of plastics can also change due to leaching processes, which in turn can affect the implications of the plastic material in the environment (Koelmanns et al. 2014).

Numerous processes cause material alteration and material degradation of plastics is slow. Half-life in nature is not only polymer dependent but depends on the environmental conditions like temperature and light exposure. For example, biodegradation rates decrease in the order polyesters > polyamides (nylon) > polyolefins (e.g., polyethylene), whereas photodegradation (alteration of materials by light) rates decrease in the order polytetrafluoroethylene > polyesters > polyamides (Arp et al. 2021).

Estimations show that for high-density polyethylene half-lives in marine environments lay between 58 years and 1,200 years (Chamas et al. 2020).

With a continuous release of plastics into the environment, the accumulation of this slowly degrading material is likely to continue.

Effects of Microplastics on Aquatic Biota

Images of aquatic organisms entangled in larger plastic products, and for this reason restricted in their mobility or even perished, are well known especially from marine habitats (e.g., Gregory 2009). Effects from microplastics are less obvious but it is unanimously accepted that the ubiquitous presence of microscopic plastic particles threatens aquatic habitats and alluvial wildlife.

Due to the high diversity of microplastic types (Fig. 1), possible effects on organisms are manifold and depend on the specific properties of the particles, such as size, shape, density, and chemical composition including plastic chemical additives (Li et al. 2018; Zimmermann et al. 2020). In addition, environmental conditions and specific characteristics of organisms play a role. For an overview and orientation, Figure 6 categorizes potential effects of microplastic particles on ecosystems and organisms. It is a simplified scheme of the highly complex interactions between particle properties and abiotic and biotic parameters. Both, the particle properties (e.g., plastic type, surface, etc.) and the environmental conditions (e.g., weathering, presence of certain micro biocenosis) determine which chemicals or microorganisms (or pathogens) adhere to particles, transport pathways or availability to various organisms. Here, we will summarize the state of research on most important isolated effects while the complexity of interactions is beyond this chapter.

In general, effects of microplastic particles can be observed at all biological levels—from subcellular to population level. Microplastic particles can affect organisms indirectly by changing their habitats (and thereby causing e.g. behavioral changes), or directly through mechanical injuries or toxicological effects. A rough distinction in the impacts on organisms can be made with regard to exposure via external contact vs. ingestion into organisms (with subsequent (eco)toxicological effects).

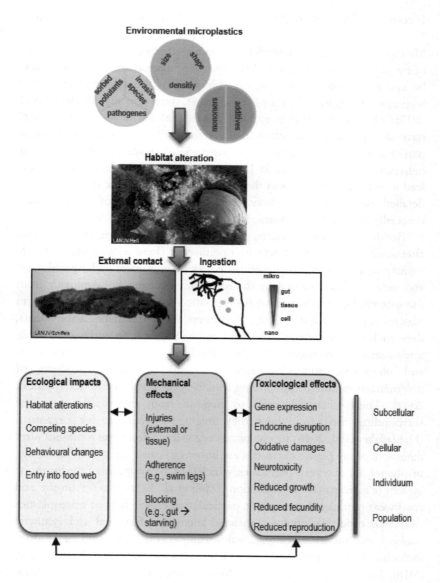

Fig. 6: *Examples of contact routes and possible effects of microplastics on ecosystems and organisms. (Source: own representation)*

Ingestion of Microplastic Particles—Potential (Eco)toxicological Effects

Microplastics research mainly focuses on the effects caused by the ingestion of microplastics in organisms. The impacts of microplastics are assumed to be caused by physical effects, e.g., blocking of the intestinal tract after ingestion leading to malnutrition or, at worst, starvation (e.g., Besseling et al. 2013; Cole et al. 2013; Gregory 2009; Wright et al. 2013). Nasser et al. (2016) have shown that accumulation of polystyrene (PS) micro- and nanoplastic particles in the gut lumen and insufficient clearance affects feeding behaviour/nutrition uptake in *Daphnia magna*. Sharp-edged particles could lead to mechanical injuries in the intestinal tract. Wright et al. (2013) give a detailed review of the physical/mechanical impacts of microplastics, especially on marine organisms, for example.

Besides physical impacts, microplastics are suggested to cause chemical/toxicological effects in organisms. These effects may include toxicity of the chemicals present in plastics (monomers and/or additives) and indirect effects caused by the sorption of organic contaminants from the surrounding medium onto microplastic surfaces. Chemicals present in plastics comprise additives, e.g., plasticizers, flame retardants, antioxidants, dyes such as nonylphenol, and antimicrobial agents. In addition to these intentionally added chemicals, other substances such as residual monomers and oligomers, impurities, and side- and breakdown products of polymerization and compounding, can be contained in plastics (Muncke 2009). These chemicals tend to migrate from the plastics into the surrounding medium, where they might be available for organisms (Hahladakis et al. 2018) or chemicals might leach from microplastics after ingestion. Many plastic chemicals are known to cause hazards in biota, e.g., bisphenol A and several phthalates, which lead to disruption of the hormone system (e.g., impaired reproduction) (Meli et al. 2020). Studies on tire and road wear particles that display a particular emerging field of microplastics deal with especially complex material mixtures of natural and synthetic rubbers, various additives (e.g., vulcanizing agents), road and brake dust, including heavy metals and other chemicals spilt on roads (Wagner et al. 2018). For example, Tian et al. (2021) identified the metabolite 6-PPD-quinone of the globally used tire additive 6-PPD as highly acutely toxic to coho salmon (*Oncorhynchus kisutche*; median lethal concentration of 0.8 ± 0.16 $\mu g/L$).

Additionally, microplastic particles are known to display a high surface area for organic contaminants to sorb at (analogously to natural occurring suspended organic matter) (Fang Wang et al. 2020; Hartmann et al. 2017), which may lead to concentration factors of organic contaminants on microplastic surfaces up to 10^6 compared to the surrounding water (Ogata et al. 2009). Wang et al. (2018) reviewed available data on microplastic interactions with chemicals in marine environments. It seems convenient that such data, at least in significant parts, also applies to freshwater habitats. Thus, microplastic particles are hypothesized to play a role as vectors for organic contaminants (or pathogens, see below) into organism tissues (i.e., trojan horse hypothesis) (Zhang and Xu 2020; Hartmann et al. 2017). Studies have shown that organisms incorporating microplastics, which were spiked with organic contaminants, accumulate these chemicals under controlled laboratory conditions (e.g., Ma et al. 2016). These studies have limited potential to be transferred to the ecosystem, as concentrations bioaccumulated by natural prey exceed the flux of ingested microplastics in most habitats (Koelmans et al. 2016). However, nanoplastics have been identified as more relevant regarding the vector function of plastics than microplastics due to their smaller size, which not only implies a higher relative surface area but also enables subcellular uptake/phagocytosis (Rehse et al. 2018; Chen et al. 2017).

Limits and Challenges in Ecotoxicology

Several research groups reviewed available effect data over the past years and concluded that there are indications for adverse effects caused by micro- or nanoplastic particle exposure (Triebskorn et al. 2019; Wright et al. 2013; Hirt and Body-Malapel 2020; Zhang and Xu 2020; de Ruijter et al. 2020; Wagner et al. 2018; Wagner and Lambert 2018; Li et al. 2018; Anbumani and Kakkar 2018). However, a returning and strong criticism of studies examining the adverse effects of microplastics is the lack of harmonization within the microplastic research community and the absence of environmentally relevant exposure scenarios (e.g., Triebskorn et al. 2019; Wagner and Lambert 2018). For instance, Triebskorn et al. (2019) evaluated 67 effect studies, with only three publications using environmentally relevant exposure scenarios: Ziajahromi et al. (2018) reported the effects of polyethylene (PE) in sediment on mortality, growth, and emergence of

Chironomus tepperi after exposure to 500 particles (1–126 μm)/kg sediment. Two other studies did not report any effects of polystyrene divinylbenzene (PSD) plastic microspheres on fish (PSD; 100 particles (97 μm)/L on *Lates calcarifer* (Guven et al. 2018)), or of PET on gammarids (102–106 particles (10–150 μm)/L on *Gammarus pulex* (Weber et al. 2018)). Other studies investigating effects on aquatic organisms mostly used particle concentrations that widely exceeded the concentrations currently occurring in natural ecosystems (see Fig. 5), which are therefore criticized regarding their environmental relevance (Triebskorn et al. 2019). Expert groups such as the SAPEA working group, that provides independent scientific advice to the European Commission, conclude that ecological risks of microplastics are currently rare, i.e. effect concentrations obtained in the laboratory are higher than environmental concentrations (SAPEA 2019). However, due to the continuing emissions of microplastics into the environment and their persistence and accumulation, concentrations are expected to increase and risks could occur within a century (SAPEA 2019).

When evaluating microplastic studies, it should also be noted that classic ecotoxicological test systems are usually used. Model organisms and test design are therefore tailored to detect the effects of homogeneously distributed, dissolved substances—but not of particulate stressors. With certain test configurations it is sometimes not even possible to ensure sufficient contact between the particle and the test organism (e.g., particles float on the surface while the test organisms are only in the water column). Additionally, it is not yet clear whether the endpoints recorded in ecotoxicological test systems are suitable and sensitive at all for detecting the effects of particulate microplastics. Thus, ecotoxicology faces new challenges in the development of adequate test strategies.

Adaption of Ecotoxicological Test Strategies

Considering the (1) need for harmonized testing strategies, (2) microplastics as a novel, anthropogenically introduced environmental matrix, and (3) the fact that not only chemical but also mechanical stress are known modes of action, no "one test for one endpoint" principle can address the complexity of this research area. A multi-dimensional testing strategy is required, which considers substance parameters (e.g., polymer type, size, and shape), single-contaminant or mixture toxicity (e.g., known virgin particles vs. unknown mixture of environmental particles), and the choice of the corresponding endpoint and the resulting suitable test method (Tab. 2, Fig. 7). For certain questions that focus primarily on toxicological effects, it may be more appropriate to test extracts (or leachates) rather than the entire particle while other questions may specifically target the combination of mechanical and chemical stress.

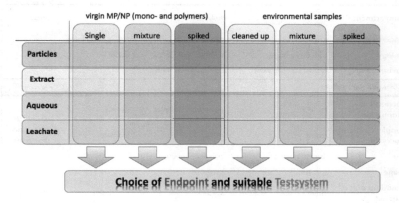

Fig. 7: Matrix schema of possible common sample types of micro- or nanoplastics. (Source: own representation)

A bioanalytical toolbox with a broad battery of effect-based methods designed to apply to toxicity testing of microplastic matrices would have to consider all the previously mentioned variables to appreciate the whole complexity of this novel anthropogenic matrix. The choice of the individual test setup depends on the physico-chemical and mechanical parameters of the investigated particles, e.g., the bioavailability of the substances of interest (e.g., buoyant particles are not available for sediment-bound test organisms). Depending on the research hypothesis, adequate model organisms should be selected that provide clear and sensitive endpoints for the effects to be expected.

An exemplary bioassay battery to analyze microplastics in the environment may comprise (1) the *in vivo* "aquatic trias" in algae, daphnids, and fish embryos, (2) three or more *in vitro* assays covering mechanistic toxicological endpoints, and (3) behavioral or reproductive endpoints or molecular biological investigations on transcript level—depending on whether the research questions is rather a biological or mechanistic point of view—to support the *in vivo* information with physiological reaction patterns.

Tab. 2: Overview of parameters that can and should be reported to improve the reproducibility of reported microplastic research. (Source: own representation)

Matrix Parameters				
Compartment	aquatic (marine/freshwater)	soil/sediment	air	urban
Processing	native	extract		
Biology	with eco-corona	without eco-corona	*in vivo*	*ex vivo*
Environment	natural environment	artificial environment		
Substance parameters				
Chemical information	purity	polymer/monomer	density	
Sample processing	native	extract	leachate	enzymatic digestion
Size	diameter/length	size distribution		
Shape	spherical	irregular		
Condition of test material	virgin product	aged	metabolite	
Hypothesis	polymer testing	sorption of chemicals		
Testing Parameters				
Test system	*in vivo*	*in vitro*	*in situ*	*in silico*
Species	vertebrate	invertebrate	ex vivo	
Concentration	mg/mL	mg/mg	particles/L	particles/g
Exposure medium	salinity	pH	temperature	nutrients
Test duration	acute	chronic	multi-generation	
Toxicological endpoints	(sub-)lethal	growth	behavior	mechanism-specific
Environmental relevance	concentration	size distribution	natural particles	

A primary reason for the overall lack of evidence and knowledge, which prevents tentative conclusions, is the complete lack of standardization in microplastic research, including a generally accepted definition. Secondly, the complexity of microplastic matrices in the environment and the currently available analytical detection limits make it challenging to conclude data from toxicity tests. Wardman et al. (2021) emphasized enhancing the efforts in developing a sound and harmonized communication strategy of microplastic toxicity research to enable the development of an evidence-based microplastic regulation strategy. To safeguard reproducibility of microplastic toxicity testing, a variety of different parameters and variables has to be considered and reported (at least in the supplementary information), most of which also apply to the good practices of substance toxicity testing.

Ecological (Long-Term) Effects—Beyond the Detection Possibilities of Ecotoxicological Test Systems

The complexity of microplastic particles and the resulting potential effects on organisms lead to increasingly complex ecotoxicological test strategies that should cover as many potential effects as possible. However, ecotoxicological test systems (of course) focus on (eco-)toxicological effects—usually caused by dissolved chemical pollutants. The endpoints considered are typically not designed to capture mechanical stresses or even impacts on ecosystem-level such as habitat changes (including shifts in biocenosis/microbial community). One example of microplastic effects caused by simple, external contact with particles resulting in a range of ecological consequences is the study by Ehlers et al. (2019). They show that the common caddisfly larvae *Lepdostoma basale*, a freshwater invertebrate, actively integrate different primary and secondary microplastics in their cases. Furthermore, Ehlers et al. (2020) demonstrated in laboratory experiments that stability of *L. basale* cases is reduced by integrated microplastic particles. This may lead to a limited protective function of the cases and thus the larvae may be more vulnerable to predators resulting in decreased survival rates. This again could have consequences for the whole aquatic biocoenosis as *L. basale* is an important primary consumer. Changes in ecosystems may also be caused by invasive species (Gregory 2009), harmful algal bloom (HAB) species (Masó et al. 2003) and/or opportunistic pathogens (McCormick et al. 2014; Zettler et al. 2013), which are transported via floating plastic particles in other regions. In addition, the presence of microplastics as a novel environmental matrix with a high surface area (depending on the particle size) implies the presence of a novel type of biofilms and microbial communities growing on these particles. These microenvironments are referred to as "eco corona", shaping them also a vector for bacteria, increasing the concern about the role of nanoplastic particles in particular (Triebskorn et al. 2019; Natarajan et al. 2021).

Because of this complexity of the microplastic matrix in the environment, a comprehensive assessment must consider all possible impacts—including potential long-term consequences of ecosystem changes.

How Much Information Is Needed to Regulate Microplastics?

The complexity of microplastic particles and their multidimensional effects on organisms are leading to the development of increasingly complex ecotoxicological testing strategies. From a purely scientific point of view, each particle could be an individual multiple stressor in itself—requiring an individual testing strategy. This leads to a dilemma between scientific precision and regulatory feasibility: From a regulatory point of view, individualized and highly complex test strategies would rather prevent reasonable measures. Here, standardized and robust test procedures are a prerequisite for a uniform assessment of all microplastic particles. In addition, current testing approaches focus too much on the purely ecotoxicological endpoints and usually do not capture broader ecosystem-level effects such as habitat changes (including biocoenosis/microbial community changes). In particular, more attention should be paid to long-term ecological impacts. Microplastics are extremely persistent in the environment and concentrations are steadily increasing. Effects on ecosystems that will only become relevant in decades, when critical concentrations are exceeded, must be taken into account today. Because later they are no longer reversible.

A threshold value for microplastics (and plastics in general) in the environment must therefore go beyond a "conventional limit value" and be based primarily on the high persistence and the unpredictable ecological consequences. This is suggested, for example, by the European Chemicals Agency proposal for a restriction on intentionally added microplastics (ECHA 2019).

Complexity of the Microplastics Problem—A Challenge for Risk Assessment and Regulation

Studies looking into the environmental fate and hazard of micro- and nanoplastics are little comparable with respect to their results, which stems from the lack of internationally accepted definition and the application of a variety of non-standardized methods, but is also due to the heterogeneity of the material and the resulting complexity in the analysis. This is one of the key bottlenecks for generating a consistent and reliable hazard assessment. To overcome this, the use of standardized methods would provide a little more clarity. While the standardization process for analytical methods is in good progress, available standard methods in ecotoxicology are typically made for dissolved compounds (Petersen et al. 2022). Micro- and nanoplastics, like engineered nanomaterials, can also cause physical effects to organisms, which is typically trying to be avoided in standardized ecotoxicology methods. Given their common particulate nature, the concepts developed for the environmental risk assessment of engineered nanomaterials may be adapted for the fate and hazard assessment of micro- and particular nanoplastics (Hüffer et al 2017). One way forward may be to use OECD guidance documents 317 (OECD 2020a) and 318 (OECD 2020b)) and test guideline 318 (OECD 2020c), which have been developed for engineered nanomaterials. Although many of the challenges in designing particles-specific tests for engineered nanomaterials are applicable to micro- and nanoplastics, it must be pointed out that there are considerations specific to micro- and nanoplastics requiring adaptations to the standard methods.

The rather low intrinsic toxicity of microplastic particles determined in the studies available to date and the results of ecotoxicological risk assessments do not hint at a widespread risk of microplastics to the environment. Due to complexity of the material and the great uncertainty regarding environmental effects, regulatory measures are being implemented in various countries around the world to reduce microplastic inputs. Although recent policy decisions, in particular the banning of microbeads in some countries, can be seen as precautionary measures, the relevant legal documents do not refer to the precautionary principle. For instance, Canada listed microbeads as toxic substances under the Canadian Environmental Protection Act (CEPA), although scientific studies indicate a rather low toxicity of microplastic particles compared to other environmental pollutants. Due to this incongruence of ecotoxicological test results and rationales used for regulatory measures, there are voices calling for a revision of the scientific rationale for microplastics regulation (Bujnicki et al. 2019).

In Europe, where a restriction of microplastics is currently being prepared by ECHA, other criteria than detrimental ecotoxicological effects are used as the basis for regulation. In this context, a qualitative risk assessment is proposed, which is generally conducted under REACH for substances for which the quantitative approach (PEC/PNEC ratio) cannot be performed with sufficient reliability. The qualitative risk assessment covers substances that satisfy PBT (persistent, bioaccumulative, and toxic) or vPvB (very persistent, very bioaccumulative) criteria and that might not show effects in short-term toxicity tests. According to ECHA, the risks of microplastics are not adequately controlled because the particles persist in the environment and each release contributes to a growing environmental stock that could exceed safe thresholds in the future (ECHA 2019). Therefore, it is proposed to treat microplastics as a non-threshold substance similar to PBT/vPvB substances under the REACH regulation. In this approach, emissions are used as a proxy for the associated risks and efforts to reduce emissions can be equated with a reduction in risk (ECHA 2019). Although microplastics have not been shown to be bioaccumulative or toxic, the ECHA approach is a way forward as it focuses on other criteria than toxicological thresholds. In this way, ECHA complies with the precautionary principle, as long-term effects on ecosystems (e.g., habitat changes) are even more difficult to assess than the immediate ecotoxicological effects on organisms. However, it is difficult to translate this approach into concrete regulatory requirements. Microplastics as an environmental problem are not subsumable as an indicator that could be used to derive the specific need for measures. In addition, much of the microplastics originates from fragmentation of larger products in the environment, i.e., after the items have left their actual product cycle. This makes it difficult to target measures to those actually responsible.

References

Anger, Philipp M., Elisabeth von der Esch, Thomas Baumann, Martin Elsner, Reinhard Niessner, and Natalia P. Ivleva (2018). Raman microspectroscopy as a tool for microplastic particle analysis. *Trends in Analytical Chemistry*, 109, 214–226.

Arp, Hans Peter H., Dana Kühnel, Christoph Rummel, Matthew Macleod, Annegret Potthoff, Sophia Reichelt, Elisa Rojo-Nieto, Mechthild Schmitt-Jansen, Johanna Sonnenberg, Erik Toorman, and Annika Jahnke (2021). Weathering plastics as a planetary boundary threat: Exposure, fate, and hazards. *Environmental Science & Technology*, 55, 7246–7255.

Arthur, Courtney, Joel E. Baker, and Holly A. Bamford (2009). Proceedings of the international research workshop on the occurrence, effects and fate of microplastic marine debris, Sept 9–11, 2008; National Oceanic and Atmospheric Administration.

Bänsch-Baltruschat, Beate, Birgit Kocher, Friederike Stock, and Georg Reifferscheid (2020). Tyre and road wear particles (TRWP) – A review of generation, properties, emissions, human health risk, ecotoxicity, and fate in the environment. *Science of the Total Environment*, 733, 137823.

Bellasi, Arianna, Gilberto Binda, Andrea Pozzi, Silvia Galafassi, Pietro Volta, and Roberta Bettinetti, (2020). Microplastic contamination in freshwater environments: a review, focusing on interactions with sediments and benthic organisms. *Environments*, 7 (30), 1–28.

Bertling, Jürgen, Ralf Bertling, and Leandra Hamann (2018). Kunststoffe in der Umwelt: Mikro- und Makroplastik. Ursachen, Mengen, Umweltschicksale, Wirkungen, Lösungsansätze, Empfehlungen. Kurzfassung der Konsortialstudie, Fraunhofer-Institut für Umwelt-, Sicherheits- und Energietechnik Umsicht, Oberhausen.

Besseling, Ellen, Anna Wegner, Edwin M. Foekema, Martine J. van den Heuvel-Greve, and Albert A. Koelmans (2013). Effects of microplastic on fitness and PCB bioaccumulation by the lugworm *Arenicola marina* (L.). *Environmental Science & Technology*, 47 (1), 593–600.

Braun, Ulrike, Korinna Altmann, Claus G. Bannick, Roland Becker, Hajo Ritter, Mathias Bochow, Georg Dierkes, Kristina Enders, Kyriakos A. Eslahian, Dieter Fischer, Corinna Földi, Monika Fuchs, Gunnar Gerdts, Christian Hagendorf, Claudia Heller, Natalia P. Ivleva, Martin Jekel, Jutta Kerpen, Franziska Klaeger, Oliver Knoop, Matthias Labrenz, Christian Laforsch, Nathan Obermaier, Sebastian Primpke, Jens Reiber, Susanne Richter, Mathias Ricking, Barbara Scholz-Böttcher, Friederike Stock, Stephan Wagner, Katrin Wendt-Potthoff, and Nicole Zumbülte (2020). Analysis of microplastics - Sampling, preparation and detection methods, Status Report. 13.04.2022 https://bmbf-plastik.de/sites/default/files/2020-1/Statuspapier_Mikroplastik%20Analytik_Plastik%20in%20der%20Umwelt_2020.pdf.

Brennholt, Nicole, Maren Heß, and Georg Reifferscheid (2018). Freshwater microplastics: Challenges for regulation and management. In Martin Wagner and Scott Lambert (eds.). *Freshwater microplastics*, 239–272. The Handbook of Environmental Chemistry, Vol 58. Cham: Springer.

Bujnicki Janusz, Pearl Dykstra, Elvira Fortunato, Nicole Grobert, Rolf-Dieter Heuer, Carina Keskitalo, and Paul Nurse (2019). Environmental and health risks of microplastic pollution. European Commission: Brussels, Belgium.

Carpenter, Edward J., and K. L. Smith Jr. (1972). Plastics on the Sargasso Sea surface. *Science*, 175 (4027), 1240–1241.

Chamas, Ali, Hyunjin Moon, Jiajia Zheng, Yang Qiu, Tarnuma Tabassum, Jun H. Jang, Mahdi Abu-Omar, Susannah L. Scott, and Sangwon Suh (2020). Degradation rates of plastics in the environment. *ACS Sustainable Chemistry & Engineering*, 8, 3494–3511.

Cole, Matthew, Pennie Lindeque, Elaine Fileman, Claudia Halsband, Rhys Goodhead, Julian Moger, and Tamara S. Galloway (2013). Microplastic ingestion by zooplankton. *Environmental Science & Technology*, 47 (12), 6646–6655.

Colton Jr., John B., Bruce R. Burns, and Frederick D. Knapp (1974). Plastic particles in surface waters of the Northwestern Atlantic: The abundance, distribution, source, and significance of various types of plastics are discussed. *Science*, 185 (4150), 491–497.

Dris, Rachid, Johnny Gasperi, and Bruno Tassin (2018). Sources and fate of microplastics in urban areas: a focus on Paris megacity. In Martin Wagner and Scott Lambert (eds.). *Freshwater microplastics*, 69–83. The Handbook of Environmental Chemistry, Vol 58. Cham: Springer.

Dümichen, Erik, Paul Eisentraut, Claus G. Bannick, Anne-Kathrin Barthel, Rainer Senz, and Ulrike Braun (2017). Fast identification of microplastics in complex environmental samples by a thermal degradation method. *Chemosphere*, 174, 572–584.

ECHA (2019). Annex XV restriction report. Proposal for a restriction: Intentionally added microplastics, European Chemicals Agency, Annankatu 18, PO BOX 400, FI-00121, Helsinki, Finland.

Ehlers, Sonja M., Werner Manz, and Jochen H. E. Koop (2019). Microplastic of different characteristics are incorporated into the larval cases of the feshwater caddisfly *Lepidostoma basale*. *Aquatic Biology*, 28, 67–77.

Ehlers, Sonja M., Tamara Al Najjar, Thomas Taupp, and Jochen H. E. Koop (2020). PVC and PET microplastics in caddisfly (*Lepidostoma basale*) cases reduce case stability. *Environmental Science and Pollution Research*, 27, 22380–22389.

Fischer, Marten, and Barbara M. Scholz-Böttcher (2017). Simultaneous trace identification and quantification of common types of microplastics in environmental samples by pyrolysis-gas chromatography–mass spectrometry. *Environmental Science & Technology*, 51 (9), 5052–5060.

Fries, Elke, Jens H. Dekiff, Jana Willmeyer, Marie-Theres Nuelle, Martin Ebert, and Dominique Remy (2013). Identification of polymer types and additives in marine microplastic particles using pyrolysis-GC/MS and scanning electron microscopy. *Environmental Science-Processes & Impacts*, 15 (10), 1949–1956.

Gregory, Murray R. (2009). Environmental implications of plastic debris in marine settings - entanglemet, ingestion, smothering, hangers-on, hitch-hiking and alien invasions. *Philosophical Transactions of the Royal Society B: Biological Sciences*, 364 (1526), 2013–2025.

Guven, Olgac, Lis Bach, Peter Munk, Khuong V.Dinh, Patrizio Mariani, and Torkel G. Nielsen (2018). Microplastic does not magnify the acute effect of pah pyrene on predatory performance of a tropical fish (*Lates calcarifer*). *Aquatic Toxicology* 198, 287–93.

Hahladakis, John N., Costas A.Velis, Roland Weber, Eleni Iacovidou, and Phil Purnell (2018). An overview of chemical additives present in plastics: Migration, release, fate and environmental impact during their use, disposal and recycling. *Journal of Hazardous Materials*, 344, 179–199.

Hartmann, Nanna B., Sinja Rist, Julia Bodin, Louise H. S. Jensen, Stine N. Schmidt, Philipp Mayer, Anders Meibom, and Anders Baun (2017). Microplastics as vectors for environmental contaminants: exploring sorption, desorption, and transfer to biota. *Integrated Environmental Assessment and Management*, 13 (3), 488–493.

Hartmann, Nanna B., Thorsten Hüffer, Richard C. Thompson, Martin Hassellöv, Anja Verschoor, Anders E. Daugaard, Sinja Rist, Therese Karlsson, Nicole Brennholt, Matthew Cole, Maria P. Herrling, Maren C. Hess, Natalia P.Ivleva, Amy L. Lusher, and Martin Wagner (2019). Are we speaking the same language? Recommendations for a definition and categorization framework for plastic debris. *Environmental Science & Technology*, 53 (3), 1039–1047.

Hassellöv, Martin, Kaegi, Ralf (2009). Analysis and characterization of manufactured nanoparticles in aquatic environments. In Jamie R. Lead and Emma Smith (eds.). *Environmental and human health impacts of nanotechnology*, 211–266. West Sussex: Wiley-Blackwell.

Hidalgo-Ruz, Valeria, Lars Gutow, Richard C. Thompson, and Martin Thiel (2012). Microplastics in the marine environment: A review of the methods used for identification and quantification. *Environmental Science & Technology*, 46(6), 3060–3075.

Hufnagl, Benedikt, Dieter Steiner, Elisabeth Renner, Martin G. J. Löder, Christian Laforsch, and Hans Lohninger (2019). A methodology for the fast identification and monitoring of microplastics in environmental samples using random decision forest classifiers. *Analytical Methods*, 11 (17), 2277–2285.

Hufnagl, Benedikt, Michael Stibi, Heghnar Martirosyan, Ursula Wilczek, Julia N. Möller, Martin G. J. Löder, Christian Laforsch, and Hans Lohninger (2022). Computer-assisted analysis of microplastics in environmental samples based on µFTIR imaging in combination with machine learning. *Environmental Science & Technology Letters*, 9 (1), 90–95.

Hüffer, Thorsten, Antonia Praetorius, Stephan Wagner, Frank von der Kammer, and Thilo Hofmann (2017). Microplastic exposure assessment in aquatic environments: learning from similarities and differences to engineered nanoparticles. *Environmental Science & Technology*, 51 (5), 2499–2507.

Huppertsberg, Sven, Daniel Zahn, Frances Pauelsena, Thorsten Reemtsma, and Thomas P. Knepper (2020). Making waves: Water-soluble polymers in the aquatic environment: An overlooked class of synthetic polymers? *Water Research*, 181, 115931.

ISO – International Organization for Standardization (2013). Plastics – Vocabulary, ISO 472. 12.04.2022 https://www.iso.org/obp/ui/#iso:std:iso:472:ed-4:v1:en

ISO – International Organization for Standardization (2020). Plastics – Environmental aspects — State of knowledge and methodologies, ISO/TR 21960:2020.

Ivleva, Natalia P. (2021). Chemical analysis of microplastics and nanoplastics: Challenges, advanced methods and perspectives. *Chemical Reviews*, 121 (19), 11886–11936.

Käppler, Andrea, Frank Windrich, Martin G. J. Löder, Mikhail Malanin, Dieter Fischer, Matthias Labrenz, Klaus-Jochen Eichhorn, and Brigitte Voit (2015). Identification of microplastics by FTIR and Raman microscopy: a novel silicon filter substrate opens the important spectral range below 1300 cm−1 for FTIR transmission measurements. *Analytical and Bioanalytical Chemistry*, 407 (22), 6791–6801.

Kawecki, Delphine, and Bernd Nowack (2019). Polymer-specific modeling of the environmental emissions of seven commodity plastics as macro- and microplastics. *Environmental Science & Technology*, 53, 9664–9679.

Koelmans, Albert A., Ellen Besseling, and Edwin M. Foekema (2014). Leaching of plastic additives to marine organisms. *Environmental Pollution*, 187, 49–54.

Koelmans, Albert A., Adil Bakir, G. Allen Burton, and Colin R. Janssen (2016). Microplastic as a vector for chemicals in the aquatic environment: critical review and model-supported reinterpretation of empirical studies. *Environmental Science & Technology*, 50 (7), 3315–3326.

Lassen, Carsten, Steffen F. Hansen, Kerstin Magnusson, Fredrik Norén, Nanna B. Hartmann, Pernille R. Jensen, Torkel G. Nielsen, and Anna Brinch (2015). Microplastics – Occurrence, effects and sources of releases to the environment in Denmark. 12.042022 https://www.tilogaard.dk/Miljostyrelsens_rapport_om_mikroplast_978-87-93352-80-3.pdf.

Li, Jingyi, Huihui Liu, and J. Paul Chen (2018). Microplastics in freshwater systems: A review on occurrence, environmental effects, and methods for microplastics detection. *Water Research*, 137, 362–374.

Liedermann, Marcel, Philipp Gmeiner, Sebastian Pessenlehner, Marlene Haimann, Philipp Hohenblum, and Helmut Habersack (2018). A methodology for measuring microplastic transport in large or medium rivers. *Water* 10 (4), 414.

Löder, Martin G. J., and Gunnar Gerdts (2015). Methodology used for the detection and identification of microplastics — A critical appraisal. In Melanie Bergmann, Lars Gutow, and Michael Klages (eds.). *Marine anthropogenic litter*, 201–227. Cham: Springer International Publishing.

Löder, Martin G. J., Mirco Kuczera, Svenja Mintenig, Claudia Lorenz, and Gunnar Gerdts (2015). Focal plane array detector-based micro-Fourier-transform infrared imaging for the analysis of microplastics in environmental samples. *Environmental Chemistry*, 12 (5), 563–581.

Löder, Martin G. J., Hannes K Imhof, Maike Ladehoff, Lena A. Löschel, Claudia Lorenz, Svenja Mintenig, Sarah Piehl, Sebastian Primpke, Isabella Schrank, Christian Laforsch, and Gunnar Gerdts (2017). Enzymatic purification of microplastics in environmental samples. *Environmental Science & Technology*, 51 (24), 14283–14292.

Ma, Yini, Anna Huang, Siqi Cao, Feifei Sun, Lianhong Wang, Hongyan Guo, and Rong Ji (2016). Effects of nanoplastics and microplastics on toxicity, bioaccumulation, and environmental fate of phenanthrene in fresh water. *Environmental Pollution*, 219, 166–173.

MacLeod, Matthew, Hans Peter H. Arp, Mine B. Tekman, and Annika Jahnke (2021). The global threat from plastic pollution. *Science* 373, 61–65.

Masó, Mercedes, Esther Garcés, and Jordi Camp (2003). Drifting plastic debris as a potential vector for dispersing Harmful Algal Bloom (HAB) species. *Scientia Marina*, 67 (1), 107–111.

McCormick, Amanda, Timothy J. Hoellein, Sherri A. Mason, Joseph Schluep, and John J. Kelly (2014). Microplastic is an abundant and distinct microbial habitat in an urban river. *Environmental Science & Technology*, 48 (20), 11863–11871.

Meijer, Lourens J.J., Tim van Emmerik, Rudd van der Ent, Christian Schmidt, Christian, and Laurent Lebreton (2021). More than 1000 rivers account for 80% of global riverine plastic emissions into the ocean. *Science Advances*, 7, eaaz5803.

Meli, Rosaria, Anna Monnolo, Chiara Annunziata, Claudio Pirozzi, and Maria C. Ferrante (2020). Oxidative stress and BPA toxicity: An antioxidant approach for male and female reproductive dysfunction. *Antioxidants*, 9 (5), 405.

Mintenig, Svenja, M., Ivo Int-Veen, Martin G. J. Löder, Sebastian Primpke, and Gunnar Gerdts (2017). Identification of microplastic in effluents of waste water treatment plants using focal plane array-based micro-Fourier-transform infrared imaging. *Water Research*, 108, 365–372.

Mintenig, Svenja M., Martin G. J. Löder, Sebastian Primpke, and Gunnar Gerdts (2019). Low numbers of microplastics detected in drinking water from ground water sources. *Science of the Total Environment*, 648, 631–635.

Muncke, Jane (2009). Exposure to endocrine disrupting compounds via the food chain: Is packaging a relevant source? *Science of the Total Environment*, 407 (16), 4549–4559.

Nasser, Fatima, and Iseult Lynch (2016). Secreted protein eco-corona mediates uptake and impacts of polystyrene nanoparticles on *Daphnia magna*. *Journal of Proteomics*, 137, 45–51.

Natarajan, Lokeshwari, M. Annie Jenifer, and Amitava Mukherjee (2021). Eco-corona formation on the nanomaterials in the aquatic systems lessens their toxic impact: A comprehensive review. *Environmental Research*, 194, 110669.

OECD (2020a). Guidance Document 317: Guidance document on aquatic and sediment toxicological testing of nanomaterials; OECD: Paris, France.

OECD (2020b). Guidance Document 318: Guidance document for the testing of dissolution and dispersion stability of nanomaterials and the use of the data for further environmental testing and assessment strategies; OECD: Paris, France.

OECD (2020c), Test Guideline 318: Dispersion stability of nanomaterials in simulated environmental media, OECD: Paris, France.

Ogata, Yuko, Hideshige Takada, Kaoruko Mizukawa, Hisashi Hirai, Satoru Iwasa, Satoshi Endo, Yukie Mato, Mahua Saha, Keiji Okuda, Arisa Nakashima, Michio Murakami, Nico Zurcher, Ruchaya Booyatumanondo, Mohamad P. Zakaria, Le Quang Dung, Miriam Gordon, Carlos Miguez, Satoru Suzuki, Charles Moore, Hrissi K. Karapanagioti, Steven Weerts, Tim McClurg, Erick Burres, Wally Smith, Michael Van Velkenburg, Judith S. Lang, Richard C. Lang, Duane Laursen, Brenda Danner, Nickol Stewardson, and Richard C. Thompson (2009). International Pellet Watch: Global monitoring of persistent organic pollutants (POPs) in coastal waters. 1. Initial phase data on PCBs, DDTs, and HCHs. *Marine Pollution Bulletin*, 58 (10), 1437–1446.

Petersen, Elijah J., Alan J. Kennedy, Thorsten Hüffer, and Frank von der Kammer (2022). Solving familiar problems: Leveraging environmental testing methods for nanomaterials to evaluate microplastics and nanoplastics. *Nanomaterials*, 12, 1332.

Primpke, Sebastian, Richard K. Cross, Svenja M. Mintenig, Marta Simon, Alvise Vianello, Gunnar Gerdts, and Jes Vollertsen (2020). Toward the systematic identification of microplastics in the environment: Evaluation of a new independent software tool (siMPle) for spectroscopic analysis. *Applied Spectroscopy*, 74, 1127.

Rehse, Saskia, Werner Kloas, and Christiane Zarfl (2018), Microplastics reduce short-term effects of environmental contaminants. Part I: Effects of bisphenol A on freshwater zooplankton are lower in presence of polyamide particles. *International Journal of Environmental Research and Public Health*, 15 (2), 280.

Rochman, Chelsea M., Cole Brookson, Jacqueline Bikker, Natasha Djuric, Arielle Earn, Kennedy Bucci, Samantha Athey, Aimee Huntington, Hayley McIlwraith, Keenan Munno, Hannah De Frond, Anna Kolomijeca, Lisa Erdle, Jelena Grbic, Malak Bayoumi, Stephanie B. Borrelle, Tina Wu, Samantha Santoro, Larissa M. Werbowski, Xia Zhu, Rachel K. Giles, Bonnie M. Hamilton, Clara Thaysen, Ashima Kaura, Natasha Klasios, Lauren Ead, Joel Kim, Cassandra Sherlock, Annissa Ho, and Charlotte Hung (2019). Rethinking microplastics as a diverse contaminant suite. *Environmental Toxicology & Chemistry*, 38, 703–711.

Roebroek, Caspar T.J., Shaun Harrigan, Tim H. M. Van Emmerik, Calum Baugh, Dirk Eilander, Christel Prudhomme, and Florian Pappenberger (2021). Plastic in global rivers: Are floods making it worse? *Environmental Research Letters*, 16, 025003.

SAPEA, Science Advice for Policy by European Academies (2019). A scientific perspective on microplastics in nature and society, Berlin: SAPEA.

Scherer, Christian, Annkatrin Weber, Friederike Stock, Sebastijan Vurusic, Harun Egerci, Christian Kochleus, Niklas Arendt, Corinna Foeldi, Georg Dierkes, Martin Wagner, Nicole Brennholt, and Georg Reifferscheid (2020). Comparative assessment of microplastics in water and sediment of a large European river. *Science of the Total Environment*, 738, 139866.

Schymanski, Darena, Barbara E. Oßmann, Nizar Benismail, Kada Boukerma, Gerald Dallmann, Elisabeth von der Esch, Dieter Fischer, Franziska Fischer, Douglas Gilliland, Karl Glas, Thomas Hofmann, Andrea Käppler, Sílvia Lacorte, Julie Marco, Maria E. L. Rakwe, Jana Weisser, Cordula Witzig, Nicole Zumbülte, and Natalia P. Ivleva (2021). Analysis of microplastics in drinking water and other clean water samples with micro-Raman and micro-infrared spectroscopy: Minimum requirements and best practice guidelines. *Analytical and Bioanalytical Chemistry*, 413, 5969–5994.

Song, Young K., Sang H. Hong, Mi Jang, Gi M. Han, Seung W. Jung, and Won J. Shim (2017). Combined effects of UV exposure duration and mechanical abrasion on microplastic fragmentation by polymer type. *Environmental Science & Technology*, 51, 4368–4376.

Thompson, Richard C., Ylva Olsen, Richard P. Mitchell, Davis Anthony, Steven J. Rowland, Anthony W. G. John, Daniel McGonigle, and Andrea E. Russell (2004). Lost at sea: where is all the plastic? *Science*, 304 (5672), 838–838.

Tian, Zhenyu, Haoqi Zhao, Katherine T. Peter, Melissa Gonzalez, Jill Wetzel, Christopher Wu, Ximin Hu, Jasemine Prat, Emma Mudrock, Rachel Hettinger, Allan E. Cortina, Rajshree G. Biswas, Flávio V. C. Kock, Ronald Soong, Amy Jenne, Bowen Du, Fan Hou, Huan He, Rachel Lundeen, Alicia Gilbreath, Rebecca Sutton, Nathaniel L. Scholz, Jay W. Davis, Michael C. Dodd, Andre Simpson, Jenifer K.McIntyre, and Edward P.Kolodziej (2021). A ubiquitous tire rubber–derived chemical induces acute mortality in coho salmon. *Science*, 371 (6525), 185–89.

Triebskorn, Rita, Thomas Braunbeck, Tamara Grummt, Lisa Hanslik, Sven Huppertsberg, Martin Jekel, Thomas P. Knepper, Stefanie Kraisa, Xanina K. Müller, Marco Pittroff, Aki S. Ruhl, Hannah Schmiega, Christoph Schür, Claudia Strobel, Martin Wagner, Nicole Zumbülte, Heinz-R. Köhler (2019). Relevance of nano- and microplastics for freshwater ecosystems: A critical Review. *TrAC Trends in Analytical Chemistry*, 110, 375–92.

Verschoor, Anja J. (2015). Towards a definition of microplastics – Considerations for the specification of physico-chemical properties; National Institute for Public Health and the Environment (RIVM), 41.

Wagner, Martin and Scott Lambert (eds.). (2018). *Freshwater microplastics: Emerging environmental contaminants?* The Handbook of Environmental Chemistry, Vol. 58. Cham: Springer International Publishing.

Wagner, Stephan, Thorsten Hüffer, Philipp Klöckner, Maren Wehrhahn, Thilo Hofmann, and Thorsten Reemtsma (2018). Tire wear particles in the aquatic environment-a review on generation, analysis, occurrence, fate and effects. *Water Research*, 139, 83–100.

Wagner, Stephan, Philipp Klöckner, Britta Stier, Melina Römer, Bettina Seiwert, Thorsten Reemtsma, and Christian Schmidt (2019). Relationship between discharge and river plastic concentrations in a rural and an urban catchment. *Environmental Science & Technology*, 53, 10082–10091.

Wang, Fen, Charles S. Wong, Da Chen, Xingwen Lu, Fei Wang, and Eddy Y. Zeng (2018). Interaction of toxic chemicals with microplastics: a critical review. *Water Research*, 139, 208–219.

Wang, Fang, Min Zhang, Wei Sha, Yidong Wang, Huizhi Hao, Yuanyuan Dou, and Yao Li (2020). Sorption behavior and mechanisms of organic contaminants to nano and microplastics. *Molecules*, 25 (8), 1827.
https://doi.org/10.3390/molecules25081827

Weber, Annkatrin, Christian Scherer, Nicole Brennholt, Georg Reifferscheid, and Martin Wagner (2018). PET microplastics do not negatively affect the survival, development, metabolism and feeding activity of the freshwater invertebrate *Gammarus pulex*. *Environmental Pollution*, 234, 181–89.

Wolff, Sebastian, Felix Weber, Jutta Kerpen, Miriam Winklhofer, Markus Engelhart, and Luisa Barkmann (2021). Elimination of microplastics by downstream sand filters in wastewater treatment. *Water*, 13, 33.

Wright, Stephanie L., Richard C. Thompson, and Tamara S. Galloway (2013). The physical impacts of microplastics on marine organisms: a review. *Environmental Pollution*, 178, 483–92.

Zettler, Erik R., Tracy J. Mincer, and Linda A. Amaral-Zettler (2013). Life in the "plastisphere": microbial communities on plastic marine debris. *Environmental Science & Technology*, 47 (13), 7137–7146.

Zhang, Ming, and Liheng Xu (2020). Transport of micro- and nanoplastics in the environment: Trojan-horse effect for organic contaminants. *Critical Reviews in Environmental Science and Technology*, 52(5), 810–846.

Ziajahromi, Shima, Anupama Kumar, Peta A. Neale, and Frederic D. L. Leusch (2018). Environmentally relevant concentrations of polyethylene microplastics negatively impact the survival, growth and emergence of sediment-dwelling invertebrates. *Environmental Pollution*, 236, 425–31.

Zimmermann, Lisa, Sarah Göttlich, Jörg Oehlmann, Martin Wagner, and Carolin Völker (2020). What are the drivers of microplastic toxicity? Comparing the toxicity of plastic chemicals and particles to *Daphnia magna*. *Environmental Pollution*, 267, 115392.

Wardman, Toby, Albert A. Koelmans, Jacqueline Whyte, and Sabine Pahl (2021). Communicating the absence of evidence for microplastics risk: Balancing sensation and reflection. *Environment International* 150, 106116.

Risk Perception: The Case of Microplastics. A Discussion of Environmental Risk Perception Focused on the Microplastic Issue

Marcos Felipe-Rodriguez, Gisela Böhm, Rouven Doran

Introduction

Risk can be generally understood as the possibility that situations or events might lead to consequences that affect aspects of what humans value (Renn and Rohrmann 2000). Risk perception involves implicit or explicit judgements of the likelihood or uncertainty as well as the desirability or undesirability of uncertain effects, which yield some benefit or cost (Eiser 2004).[1] The formal definition of risk often entails the magnitude and probability of harmful consequences (Aven and Renn 2009), and risk perceptions include these dimensions, along with perceptions of familiarity and controllability, dread and catastrophic potential, as well as affective and emotional responses (Finucane et al. 2000; Slovic 2000; 2016). Risk perceptions deviate from numerical risk estimates because they are not exclusively determined by statistics and probabilities, but also by qualitative factors related to the risks themselves and those perceiving them (Kortenkamp and Moore 2011).

Environmental risks diverge from other types of risks. First, they are often characterized by high uncertainty and complexity, leading to complicated causal relationships and numerous consequences (Steg and De Groot 2018). Moreover, they tend to develop from the behaviors of many individuals; consequently, mitigation requires the actions of many people. Lastly, their consequences are often temporally delayed and geographically distant. Those who contribute to the risk are not necessarily those who

---

[1] Slovic (1999) argued that "danger" is a reality, but "risk" is socially constructed.

suffer its consequences, which raises ethical issues (Steg and De Groot 2018).

The current chapter focuses on risk perception of the environmental problem of microplastics. Microplastics are tiny particles of plastic, smaller than 5 mm. Microplastics are found at growing concentrations in the environment and have accumulated even in the most distant (van Sebille et al. 2015; Egger et al. 2020). The public and academia are increasingly concerned about the possible effects of this global challenge (SAPEA 2019). Accordingly, it is necessary to gain an understanding of people's perceptions and engage the public to tackle this problem effectively (Pahl and Wyles 2017).

The determinants of microplastics risk judgments are numerous and interrelated. Socio-psychological factors have substantial influence on the evaluation of such environmental risks. Furthermore, people's risk perceptions about microplastics are important to consider when addressing the threat microplastics pose. This chapter discusses the different aspects that may affect environmental risk perceptions, focusing on the case of microplastics. It begins by highlighting the characteristics of the hazard itself and moves on to the individual characteristics, with an emphasis on the role of heuristics, emotions, and finally models.

Hazard Characteristics

There are cases in which people are wary of hazards that experts agree do not cause much significant harm, like electronic radiation from mobile telephones, while in other cases, people are ready and willing to expose themselves to hazards that result in large numbers of fatalities each year, such as drinking alcohol (Siegrist and Arvai 2020). Such divergences can be at least partially explained by specific characteristics of the hazards themselves. The core variables in risk perception research are (perceived) magnitude of the risk and risk acceptance (Renn and Rohrmann 2000). Nonetheless, in most studies, many more risk-related aspects are included, such as qualitative features of the hazard (e.g., familiarity with the risk or associated fear), benefits (e.g., attractiveness of the risky activity), personal relation to the hazard (e.g., whether one voluntarily exposes oneself to it, degree of worry about the risk, etc.), and acceptability facets (e.g., willingness to pay or desired level of restrictions) (Renn and Rohrmann 2000).

The psychometric paradigm (Fischhoff et al. 1979) suggests that different types of hazards can be mapped onto four dimensions across two axes, labelled dread risk and unknown risk, respectively.[2] Dread risk refers to the level to which the risk is perceived as alarming or as having grave consequences; unknown risk describes the level to which the risk is experienced as unfamiliar, new, unobservable or having delayed effects (Steg and de Groot 2018). Dread risk includes features such as uncontrollable, catastrophic, dreaded, involuntary, fatal, inequitable, global, and difficult to reduce, whereas unknown risk includes risks that are unobservable, not understood by science, new, and have delayed effects (Kortenkamp and Moore 2011). Risks with effects that are perceived as far off in time or as occurring in a faraway place are also included within this dimension (Eyal et al. 2008). Non-experts' risk perceptions have been shown to correlate with these main dimensions. Risks rated as more dreadful and more unknown are perceived as riskier and less acceptable (Kortenkamp and Moore 2011). Other investigations have identified more dimensions that are relevant for environmental challenges, such as whether people have moral concerns related to the risk and whether people feel that issues of equity are related to the risk (e.g., Bostrom et al. 2020). Another important factor discussed in the context of risk acceptance is whether the source of the risk is natural versus human/technological, as people tend to rank natural hazards lowest in risk magnitude ratings. These hazards seem to be perceived and evaluated as more tolerable than those stemming from human activities or technologies, even though objective risk assessments might not differ much (Renn and Rohrmann 2000).

The ubiquity of microplastics in aquatic ecosystems has provoked a broad public debate on the unsustainable use and environmental impact of plastics (Kramm and Völker 2018). However, as stated above, there are cases where the public perception of a particular hazard does not match experts' understanding of its impacts. While the environmental impacts of microplastics are not at all clear from a scientific perspective at present (see Heß et al. in this volume), public awareness of overall plastic pollution is extensive (Völker et al. 2020; Kramm et al. 2022). In fact, most EU citizens worry about the consequences of plastics for the environment (87 percent) and for their own health (74 percent) (European Commission 2017). Mean-

---

2 For empirical evidence, see Slovic (1987) and Teigen et al. (1988).

while, there has been ongoing debate about the relevance of this issue compared to other environmental challenges (Backhaus and Wagner 2020), with some scholars arguing that the levels of environmental toxicity detected so far are too low to be of significant concern (Triebskorn et al. 2019). Such disparity between experts and public opinion can potentially be problematic when it results in policies and decisions that are disproportionate or not supported by science (Rist et al. 2018).

The public indeed has reported to be highly concerned that microplastics could have an impact on the sustainable development of ecosystems and also threaten food safety and public health (European Commission 2020; German Federal institute for Risk Assessment 2020; SAPEA 2019). Such levels of perceived risk might be explained in part by known/dread factors from the psychometric paradigm. With respect to the dread risk dimension, plastic and microplastics pollution are likely to be considered dreadful hazards given that plastic pollution is a form of involuntary exposure for animals and plants in the environment and a problem at a global scale. With respect to the unknown risk dimension, microplastics are a quite new hazard (Pahl and Wyles 2017) that is not well understood by science, since research on them is still in its infancy (Rist et al. 2018). This may lead people to perceive this hazard as less well understood by science. Nonetheless, 53 percent of respondents in Kramm et al. (2022) perceived the state of scientific knowledge on microplastics as rather high or very high, suggesting the contrary. Moreover, microplastics are not easily observable (Pahl and Wyles 2017), which should also make people lean more towards the unknown end of the spectrum. Additionally, regarding the sources of microplastics, since hazards stemming from human activities are perceived as riskier and less tolerable (Renn and Rohrmann 2000), the fact that microplastics are a human-caused hazard might contribute to the high levels of perceived risk that have been reported.

Perceiver Characteristics

Perceivers of risk differ on a wide range of variables that might influence risk perceptions (Siegrist and Árvai 2020). Many such variables have been studied extensively in order to explain and predict individual differences in risk perceptions. In the case of microplastics, only a few studies have been conducted to investigate public risk perceptions (Yoon et al. 2021). Perceived consequences as well as knowledge and awareness determined pro-environmental attitudes in a study by Soares et al. (2021) in Portugal. Risk perception was also a pivotal determinant of pro-environmental behavioral intention related to microplastics in Korea (Yoon et al. 2021). The relation between expectations and perception (Tsiotsou 2006) has also been found to be key in consumer decisions about green microplastics-free products (Nam et al. 2017; for a discussion on interventions to reduce plastics consumption see Grünzner and Pahl in this volume). On that note, the most relevant individual characteristics and their implications for risk perceptions of microplastics are highlighted in the sections below.

Socio-Demographics

Gender appears frequently to be weakly associated with risk perceptions (Cullen et al. 2018; Rivers et al. 2010); in addition, small or non-significant effects have been found for age (Bearth et al. 2019) as well as income (Nardi et al. 2020) and education (Bearth et al. 2019; Nardi et al. 2020). However, some studies have yielded more information about the relationship between risk perceptions and demographic characteristics. For example, in studies by Finucane and colleagues (2000), white women perceived significantly higher levels of risk across different hazards compared to white males, while the same was not found for nonwhite women and men. This indicates that gender and/or racial identity per se might not drive risk perceptions to the same extent as other psychological or cultural features (Rivers et al. 2010). There is also a notion that white males tend to have lower risk perceptions than white females, nonwhite males, and nonwhite females across different hazards (cf. white male effect; Kortenkamp and Moore 2011).

Unsurprisingly, gender effects have been found for environmental risks such as climate change (Finucane et al. 2000). Regarding microplastics, Deng et al. (2020) conducted face-to-face interviews and a structured questionnaire among residents of Shanghai (China) to investigate perceptions of microplastics, exploring willingness to reduce microplastics and its influencing factors. In this study, males had a lower average score than females on willingness to reduce microplastics emissions (Deng et al. 2020). Although differences in knowledge have been argued to be a reason for gender differences in environmental risk perceptions, females exhibited higher nuclear risk perceptions even in a sample of scientists (Barke et al. 1997). Therefore, it has been argued that there is more support for race and gender as explanations for differences in environmental risk perceptions than for differences in knowledge (Davidson et al. 1996).

Furthermore, environmental risk perception differs based on respondents' socio-economic status (Bickerstaff 2004). People with a lower social status and fewer privileges tend to be in a position of less power and control. They are argued to be more vulnerable to economic stressors and therefore perceive the world as a more dangerous place (Finucane et al. 2000). A similar trend was reported by Deng et al. (2020), who noted that people with lower education had higher levels of worry about microplastics, while people with higher education were not as concerned. The authors argued that this was due to a more comprehensive understanding of microplastics in the latter group, which in turn might have reduced unnecessary concerns. Nonetheless, another study by Henderson and Green (2020) concluded that people with high environmental awareness are also more concerned and know more about microplastics.

Knowledge and Reasoning

It is a common finding in the literature that laypeople and experts tend to differ in their level of perceived risk (Savadori et al. 2004; Siegrist et al. 2018). Sjöberg (1998) classified comparisons of experts' and laypeople's risk perceptions into three types: similar assessments for well-known risks; lower risk perceptions by laypeople for hazards which they have some control over, such as smoking or drinking; and lower risk perceptions by experts for complex topics such as nuclear power. These differences can be accounted for in part by aspects of the psychometric model (Slovic 1987), including familiarity, controllability, and knowledge. The knowledge deficit model argues that if laypeople increased their knowledge, they would reach similar conclusions to those of experts; therefore, general knowledge and risk perception should correlate (Bubela et al. 2009). In its simplest form, however, the knowledge deficit model has not garnered much empirical evidence, and there is research that casts doubt on it (Kellstedt et al. 2008).

Regarding the issue of microplastics, in the study by Deng et al. (2020), the majority of people became worried or even overly worried when informed about possible impacts of microplastics, and increased knowledge about the issue was also associated with a greater willingness to take action to tackle the problem (Deng et al. 2020). Moreover, Henderson and Green (2020) investigated people's knowledge and understanding of microplastics in the United Kingdom. Particular focus was placed on the role of the media in framing perceptions, involving participants with no knowledge of microplastics as well as participants with particular interest in microplastics. The findings shed light on the importance of environmental awareness and how lack of awareness of the plastics problem represents a barrier to change (Henderson and Green 2020). These findings highlight the importance and benefits of citizen science activities, which can raise awareness and knowledge about plastic litter (for more see Severin et al. in this volume). For instance, participation in beach clean-ups and other coastal activities has been shown to be associated with pro-environmental intentions and higher marine awareness (Wyles et al. 2017).

Recent findings by Kramm et al. (2022) showed that 80 percent of the German public had heard of microplastics, hence indicating that the public is becoming more aware of microplastics. The same investigation also found that level of education was important for microplastics awareness, since 90 percent of people considered to have a high level of education reported having heard of microplastics, whereas only 65 percent of those with low education reported having heard of them. Their results also indicated that higher environmental awareness tends to be associated with higher risk perceptions and that the more frequently one hears about microplastics, the higher the perceived risk of microplastics (Kramm et al. 2022).

According to a study by Grünzner, Pahl, White and Thompson (2021), experts (researchers working primarily on plastics) are more highly concerned about the risks of microplastics for the natural environment than they are about their risks for human health. Accordingly, microplastics have often been depicted in the media as something to be concerned about, as a risk for the environment (Völker et al. 2020). Nonetheless, some recent reports show that laypeople are highly worried about microplastics risks to the natural environment (European Commission 2020), but also quite concerned about possible health risks (German Federal Institute for Risk Assessment 2020).

People with higher levels of scientific reasoning have been found to be more likely to perceive risks consistently with the scientific evidence regarding those risks (Siegrist and Árvai 2020). Nevertheless, risk perceptions among people with high scientific reasoning ability may not correspond to the actual scientific evidence if people have already made up their minds that the hazard is of high or low risk (Drummond and Fischhoff 2019).

Fairness, Value Orientations, and Worldviews

Regarding perceptions of environmental risks, people tend to care less about statistics, such as the number of casualties due to a hazard, and more about issues such as justice, fairness, and duties to future generations (Moore 2009). Within the psychometric model, the dread component of risk contains aspects related to ethical issues resulting from the unequal distribution of and lack of informed consent regarding risk exposure (Slovic 1987). Additionally, moral evaluations of risks have proven to be a strong predictor of acceptability and perceived risk (e.g., Sjöberg and Drottz-Sjöberg 2001); likewise, environmental injustice has been found to predict risk perceptions (Satterfield et al. 2004).

Cultural worldviews are defined as the pattern of beliefs and value orientations shared by people in a collective, or orienting inclinations which guide thoughts and behaviors (Mead and Métraux 1954). Such worldviews are argued to have a strong influence on risk perceptions. Individuals and collectives ascribe to one or a set of prominent value orientations, namely hierarchism, individualism or egalitarianism (cf. cultural theory of risk; Douglas and Wildawsky 1982). Later research expanded value orientations to include egoism, altruism, and most interestingly for environmental risks, biospherism (e.g., De Groot and Steg 2007). These studies have pointed to a weak relationship between worldviews and risk perceptions overall, albeit with two particular environmental hazards as noteworthy exceptions: nuclear power and climate change (Siegrist and Arvai 2020). On another note, some people hold beliefs that lack scientific basis, such as so-called "New Age beliefs" (Sjöberg and Wahlberg 2002). Sjöberg and Wahlberg (2002) investigated risk perception in relation to these beliefs, including traditional folk superstition, belief in paranormal phenomena and use of alternative healing practices. Such beliefs explained 15 percent of the variance in perceived risk (Sjöberg and Wahlberg 2002). People with such beliefs tend to hold higher risk perceptions regarding environmental hazards such as climate change and nuclear waste (Siegrist and Arvai 2020).

There are no studies on perceptions of microplastics that have explored value orientations or worldviews. Nevertheless, in studies of the risk of nuclear power, altruistic and biospheric values tended to be negatively associated with perceived risks (Siegrist and Arvai 2020). For climate change, on the contrary, biospherism, and to a lesser extent egoism, have been positively associated with perceived risk (Van der Linden 2015). Other evidence suggests that biospheric values may partially undergird climate change worry, whilst being directly and positively related to personal climate mitigation behaviors (Bouman et al. 2020).

Heuristics

Prospect theory (Kahnemann and Tversky 1979) postulates that people tend to overweight small probabilities and underweight larger probabilities, depending on the type of decision they are making. Specifically, people overweight small probabilities when simply presented with descriptions of these probabilities, yet tend to underweight small probabilities when they are learned through experience (Kahnemann and Tversky 1979). It is important to mention that people often lack the in-depth knowledge needed to evaluate hazards comprehensively, as indicated by studies addressing technologies (Connor and Siegrist 2011) and climate change (Shi et al. 2016).

The elaboration likelihood model (Petty and Cacioppo 1986; for an application in the environmental context, see Meijnders et al. 2001) argues that lack of motivation or knowledge leads to the usage of a peripheral cognitive route, where heuristics are prominent. Heuristics are argued to work through attribute substitution (Kahneman and Frederick 2005). When evaluating a hazard, an attribute that is not cognitively accessible, such as the probability of being exposed to the hazard, it is substituted with an attribute that is more easily accessed, such as recollection of concrete examples of that hazard (Siegrist and Arvai 2020). For example, someone is more likely to evaluate the hazard of plastic pollution based on number of the times they spotted plastic floating in the sea rather than the actual statistical probability of exposure to plastic pollution.

Availability Heuristic

The availability heuristic is used when people utilize the "ease" with which examples or occurrences can be brought to mind to assess the probability of an event (Tversky and Kahneman 1974). For instance, someone might assess the risk of microplastics negatively affecting the environment by thinking about how often they hear in the news that microplastics have been found in their local area. The availability heuristic has been examined with respect to environmental hazards such as flooding (e.g., Tanner and Arvai 2018), with people who could remember floods perceiving higher risk compared to those who could not remember such events. One might speculate that the use of this heuristic might similarly affect laypeople's perceived risk when it comes to microplastics. Support is provided by literature indicating that this heuristic may influence risk perception regarding climate change (Demski et al. 2017).

Affect Heuristic

The affect heuristic maintains that the affective component elicited by a hazard influences risk perception (Finucane et al. 2000). People are argued to base their judgements about risks and benefits on their affective reactions (Slovic 1999). It is further argued that there is an affect "pool" that contains positive and negative markers associated with all mental images (Slovic et al. 2004). Studies investigating this principle suggest that the valence of spontaneous associations is associated with risk perceptions and acceptance of risk (Siegrist and Arvai 2020). The problem is that the affect heuristic might result in biased judgements (Nakayachi 2013) by leading people to ignore information that would have been useful to formulate more accurate risk judgements (Sunstein 2003). Accordingly, one possible explanation for the fact that people have been reported to perceive microplastics as more harmful than what the scientific evidence appears to indicate at this stage (Catarino et al. 2021) could be that people might associate microplastics with a negative affective component. Thus, people might be biased to think negatively about microplastics impacts and ignore certain information, in this case current uncertainty about the impacts, particularly on human health. Furthermore, questions about the causal direction of these associations can be posed. It is hard to exclude the possibility that risk perception might drive affective responses and not the other way around (Siegrist and Arvai 2020).

Natural-Is-Better Heuristic

In Western countries, nature is generally perceived as benevolent (Scott and Rozin 2020). The natural-is-better heuristic is defined as neglecting the positive effects of human intervention and negative impacts of natural processes (Siegrist and Hartmann 2020). Research in this vein shows that synthetic chemicals are much more negatively perceived than natural chemicals (Saleh et al. 2019), especially among individuals with high biospheric values (Campbell-Arvai 2019). It follows that people might evaluate the issue of microplastics more negatively because they result from a human process, reducing naturalness. Supporting evidence stems from studies showing that microplastics are indeed perceived quite negatively (e.g., Deng et al. 2020) and that microbeads are perceived as an "unnatural", unacceptable risk (Anderson et al. 2016).

Emotions

Risk perception used to be seen as exclusively cognitive, and emotions were not considered in this field for a long time (Böhm and Brun 2008). An early study by Johnson and Tversky (1983) showed that people's current mood affected their risk judgements, highlighting that people hardly ever react to threats in an emotionally neutral state, and emotions affect how they perceive risks. Emotions have since then come to be considered important factors that affect risk perceptions and evaluations (Böhm and Tanner 2019). The previous section discussed how affect might root judgements about risks and benefits; however, there is an important distinction between general affect and specific emotions or appraisals (e.g., Lerner and Keltner 2001).

Emotions can be connected to complex reasoning; each emotion carries a certain meaning and reflects a cognitive structure or viewpoint (Böhm 2003; Böhm and Pfister 2000; 2017). For example, worry anticipates that something bad might happen in the future; outrage involves assigning blame to other people; disappointment means that an outcome has fallen short of expectations; regret arises from a sense of responsibility; pity is a social emotion; guilt is more focused on ourselves and our own actions, when we feel we have acted in a way that violates a moral norm; fear is similar to worry because it has to do with anticipating future harm, but is more short-term and intense; hope reflects the belief that there is still a chance to achieve positive outcomes; and lastly, pride is felt when something has been accomplished (Böhm 2003; Böhm and Pfister 2000; 2017). Specific emotions are often explained through the appraisal theoretical framework (Frijda 2007).

Böhm and Pfister (2000; 2017) conceptualized a dual-process model of risk evaluation involving two fundamental appraisal dimensions linked to specific emotions, namely consequences and morality. Fear is an emotion related to consequences, whereas outrage and guilt have more to do with the perception that moral norms have been violated (Böhm 2003; Böhm and Pfister 2000; 2017). Each dimension is associated with characteristic behavioral tendencies. Typical consequentialist behavioral tendencies are mitigation and adaptation, while actions more tied to morality are punishment and redemption, targeting the actor or aggressor (Böhm and Pfister 2017). A consequentialist focus could lead to judgements about the perceived risk of microplastics towards animals, which could trigger an emotion of fear and ultimately a behavioral tendency to help clean up a beach full of plastic litter.

Research indeed shows that some of the intuitive associations with microplastics tend to concern harmful impacts on wildlife (Deng et al. 2020; Henderson and Green 2020). Since the release of microplastics in the environment is evidently due to human activity, deontological evaluations might also be more intense than for natural hazards such as flooding. This interpretation follows literature showing that deontological judgments tend to be more intense in cases of human rather than natural causation, whereas consequentialist evaluations tend to be more intense when the consequences affect humans as opposed to nature (Böhm and Pfister 2000; 2017). Despite the scientific evidence being not yet clear with respect to harmful consequences of microplastics for human health (Catarino et al. 2021), the public has been repeatedly found to be worried about human health effects (Deng et al. 2020; Henderson and Green 2020). It is because of this that microplastics might actually trigger both deontological (outrage, guilt, etc.) and consequentialist emotions (sadness, fear, etc.).

Mental Models

An important basis for people's risk perceptions is how they mentally represent the risk event in question (Bostrom 2017; Böhm and Pfister 2000; 2017). Such a mental representation, commonly referred to as mental model, is constructed from available information, of which the most important components are the causes and consequences ascribed to the risk event. A person's mental model of microplastics may convey the belief that microplastics are released into aquatic environments by washing fleece and synthetic clothing and that they will result in harm to some fish species. Laypeople's mental models tend to be less structured than those of experts (Bostrom 2017). Inaccuracies in their mental models can lead people to make errors, which in the case of plastic can be seen in the development and promotion of certain actions and policies not fully supported by scientific evidence (Catarino et al. 2021). Common approaches to capture mental models about a given risk event are surveys (Bostrom 2017) as well as thought listing and image association tasks (e.g., Smith and Joffe 2013). Themes are inductively derived from open-ended responses to questions. For example, one might tap into the first things that come to people's minds when thinking of the environmental hazard; additional content analysis might refer to psychological theories (e.g., Böhm et al. 2018).

One manner to assess the utility of mental models is to employ them within a problem-solving or decision-making approach, a strategy exemplified by the mental model approach to risk communication (Morgan et al. 2002). The phases of this approach include, first, developing a conceptual model of the target system, such as microplastics, into a decision model representing how science may best inform policy and risk mitigation decisions. Hence, the conceptual model consists of decisions about risks and what could be done about them (Bostrom 2017). Second, semi-structured interviews assessing mental models and related perceptions of the risks of the issue in question and how to mitigate those risks are content analyzed and compared to the decision model (Bostrom 2017). The interview protocols often include a think-aloud task, inspired by think-aloud studies used in other mental model approaches (Ericsson and Fox 2011), but primarily consist of prompts asking participants to talk about the hazard. The analysis of the interviews is conceptually linked to the decision model, but open-ended (Bostrom 2017). Third, the interviews might inform the design of survey instruments to survey larger samples, ideally representative of the groups for whom risk communication strategies are being developed (Bostrom 2017). Another way to assess mental models regarding environmental issues is based on systems modelling and entails experiments in which individuals solve tasks such as dynamic greenhouse gas problems (e.g., Moxnes and Assuad 2012).

Regarding what people's mental representations of microplastics might look like, Dilkes-Hoffman et al. (2019) asked members of the Australian public to state the first two words that came to mind when they heard the word "plastic". The most frequent words or concepts were general environmental statements, waste, pollution, ocean impacts, and animal impacts. Participants placed the main responsibility for reducing plastic waste on industry, followed by government, and 80 percent expressed a desire to reduce their personal plastic use (Dilkes-Hoffman et al. 2019). Moreover, the study by Deng et al. (2020) in China found that microplastics seem to be viewed as accumulating mostly in the ocean. The respondents also referred to factory production of plastic particles as the main source of microplastics, although, overall, they did not seem to be fully aware of the origin of microplastics (Deng et al. 2020). In addition, a UK study by Henderson and Green (2020) reported that most respondents were unaware of microplastics, although environmentally conscious individuals had heard about microbeads through media reporting on new regulations. While some people made a connection between their personal use of plastics and ocean pollution, they appeared unable to define the link between macro- and microplastics (Henderson and Green 2020).

The few existing studies on perceptions of microplastics indicate considerable misconceptions, such as the recurrent association of microplastics with plastic islands (Henderson and Green 2020), and that an important share of the public seems to still be unaware of microplastics (Deng et al. 2020; Henderson and Green 2020)—even though microplastics awareness is on the rise (Catarino et al. 2020). It is because of this that one might speculate that laypeople's mental models of microplastics are inaccurate. Notably, laypeople most commonly associate plastic and microplastics with pollution in the environment in general and the ocean in particular (Dilkes-Hoffman et al. 2019). Laypeople seem to often fail to recognize that microplastics also migrate among the atmosphere, freshwater, soil and different creatures (Bin et al. 2020). Further research could investigate whether people are able to understand that microplastics can also be released into these environments and the impacts this could have, such as harmful effects on wildlife. Regarding human health effects, the public appears to be very concerned about these impacts, despite the fact that scientific evidence for such effects is still unclear (Catarino et al. 2020).

Plastic particles from factories are the main source of global plastic waste (Boucher and Friot 2017), and the public has reportedly made associations between these particles and microplastics (Deng et al. 2020). It could be that the majority of the public understand that these particles are the main source of microplastics, as Deng et al. (2020) argued. Nonetheless, there are various other sources of microplastics that the public generally seems not to be aware of. For instance, the decomposition of synthetic textiles is another important source of microplastics in the ocean (Boucher and Friot 2017), although the public does not make this association often (Deng et al. 2020). Additionally, the public often does not associate individual plastic consumption with the release of microplastics and thus ocean pollution (Henderson and Green 2020); instead, they often attribute responsibility to industry and government (e.g., Dilkes-Hoffman et al. 2019).

Public awareness of microplastics is increasing (Catarino et al. 2021). People are becoming more frequently exposed to the topic through the media, which is why people's awareness of the issue may continue to rise in the future. Media storytelling might indeed have a central role to play in shaping public understanding and bringing the topic to public attention in powerful ways (Henderson and Green 2020). Employing a mental model approach to risk communication may provide valuable insights into how to address the gap between experts' and laypeople's knowledge (Pahl and Wyles 2017). The mental models elicited from this research can be used to adapt messages to communicate the different risks posed by microplastics, and such communications can be evaluated via surveys or focus groups (Pahl and Wyles 2017).

Conclusions

Microplastics are a global environmental challenge that appears to increasingly concern both the public and academia. As an environmental risk, there are characteristics of microplastics as a hazard that influence people's risk perceptions. They are likely to be considered dreadful hazards given their global scale and potential impacts on animals and plants. Moreover, they are likely to be perceived as an unknown hazard, given that they are quite new, not well understood by science and not easily observable. This, together with the fact that they are caused by humans, might contribute to the high levels of perceived risk that have been reported.

Nonetheless, there are numerous relevant individual-level variables that can also shape microplastics risk perceptions. Socio-demographic characteristics, such as gender and level of education, have been shown to predict different levels of perceived risk of microplastics. Higher levels of knowledge about the issue tend to be associated with higher risk perceptions, even sometimes leading laypeople to be overly concerned. Perceivers' worldviews and values have also been reported to affect risk perceptions, among which biospherism is particularly relevant for environmental risks such as microplastics. Other potential sources of influence include the use of heuristics as opposed to more complex information processing, particularly the availability, affect, and natural-is-better heuristics.

Emotions are another important factor that affects risk perceptions, and microplastics might trigger both consequentialist and moral emotions, which in turn trigger different behavioral tendencies. Lastly, how people mentally represent microplastics in terms of, amongst other things, its causes, consequences, and possible solutions, that is, people's mental models, might also affect risk perceptions of microplastics.

Given the few studies on microplastics risk perception due to the infancy of the field, it is difficult to make solid claims about how microplastics risk perception is formed. Nonetheless, research on microplastics is increasing exponentially, including interdisciplinary projects that combine findings from the natural sciences with insights from the social and behavioral sciences. This will deepen our understanding of what determines people's perceived risk of microplastics and allow us to effectively tackle this global challenge

Acknowledgements

An earlier version of this chapter was submitted as a report for the project LimnoPlast: Microplastics in Europe Freshwater Ecosystems, from sources to solutions. This project has received funding from the European Union's Horizon 2020 research and innovation program under grant agreement No 860720. Responsibility for the information and views set out in this document lies entirely with the authors.

References

Anderson, Alison, Jane Grose, Sabine Pahl, Richard C. Thompson, and Kayleigh J. Wyles (2016). Microplastics in personal care products: Exploring perceptions of environmentalists, beauticians and students. *Marine Pollution Bulletin*, 113 (1–2), 454–460.

Aven, Terje, and Ortwin Renn (2009). On risk defined as an event where the outcome is uncertain. *Journal of Risk Research*, 12 (1), 1–11.

Backhaus, Thomas, and Martin Wagner (2020). Microplastics in the environment: Much ado about nothing? A debate. *Global Challenges*, 4 (6), 1900022.

Barke, Richard P., Hank Jenkins-Smith, and Paul Slovic (1997). Risk perceptions of men and women scientists. *Social Science Quarterly*, 78 (1), 167–176.

Bearth, Angela, Rita Saleh, and Michael Siegrist (2019). Lay people's knowledge about toxicology and its principles in eight European countries. *Food and Chemical toxicology*, 131, 110560.

Bickerstaff, Karen (2004). Risk perception research: socio-cultural perspectives on the public experience of air pollution. *Environment International*, 30 (6), 827–840.

Bin, Zhao, Cheng Yongqiang, Guo Cuilian, Liu Maoke, Yao Puyu, and Zhao Yang (2020). Outlook and overview of microplastics pollution in ecological environment. In *E3S Web of Conferences* (Vol. 143, p. 02027). Les Ulis. *EDP Sciences*. https://doi.org/10.1051/e3conf/202014302027

Böhm, Gisela (2003). Emotional reactions to environmental risks: Consequentialist versus ethical evaluation. *Journal of Environmental Psychology*, 23 (2), 199–212.

Böhm, Gisela, and Wibecke Brun (2008). Intuition and affect in risk perception and decision making. *Judgment and Decision Making*, 3, 1–4.

Böhm, Gisela, Rouven Doran, and Hans-Rüdiger Pfister (2018). Laypeople's affective images of energy transition pathways. *Frontiers in Psychology*, 9, 1904.

Böhm, Gisela, and Hans-Rüdiger Pfister (2000). Action tendencies and characteristics of environmental risks. *Acta Psychologica*, 104 (3), 317–337.

Böhm, Gisela, and Hans-Rüdiger Pfister (2017). The perceiver's social role and a risk's causal structure as determinants of environmental risk evaluation. *Journal of Risk Research*, 20 (6), 732–759.

Böhm, Gisela, and Carmen Tanner (2019). Environmental risk perception. In Linda Steg and Judith I. M. de Groot (eds.). *Environmental psychology: An introduction*, 13–25. Hoboken, NJ: Wiley.

Bostrom, Ann (2017). Mental models and risk perceptions related to climate change. In Ann Bostrom (ed.). *Oxford research encyclopedia of climate science*. Oxford: Oxford University Press.

Bostrom, Ann, Gisela Böhm, Adam L. Hayes, and Robert E. O'Connor (2020). Credible threat: perceptions of pandemic coronavirus, climate change and the morality and management of global risks. *Frontiers in Psychology*, 11, 578562.

Boucher, Julien, and Damien Friot (2017). *Primary microplastics in the oceans: a global evaluation of sources* (Vol. 10). Gland, Switzerland: Iucn.

Bouman, Thijs, Mark Verschoor, Casper J. Albers, Gisela Böhm, Stephen D. Fisher, Wouter Poortinga, Lorraine Whitmarsh, and Linda Steg (2020). When worry about climate change leads to climate action: How values, worry and personal responsibility relate to various climate actions. *Global Environmental Change*, 62, 102061.

Bruine de Bruin, Wändi, and Ann Bostrom (2013). Assessing what to address in science communication. *Proceedings of the National Academy of Sciences*, 110 (Supplement_3), 14062–14068.

Bubela, Tania, Matthew C. Nisbet, Rick Borchelt, Fern Brunger, Cristine Critchley, Edna Einsiedel, Gail Geller, Anil Gupta, Jürgen Hampel, Robyn Hyde-Lay, Eric W. Jandciu, Ashley Jones, Pam Kopopack, Summer Lane, Tim Lougheed, Brigitte Nerlich, Ubaka Ogbogu, Kathleen O'Riordan, Colin Ouellette, Mike Spear, Stephen Strauss, Thushaanthini Thavaratnam, Lisa Willemse and Timothy Caulfield (2009). Science communication reconsidered. *Nature Biotechnology*, 27 (6), 514–518.

Campbell-Arvai, Victoria (2019). Visits from the ghost of disturbance past: Information about past disturbance influences lay judgments of ecosystems. *Journal of Environmental Management*, 232, 438–444.

Catarino, Anna I., Johanna Kramm, Carolin Völker, Theodore B. Henry, and Gert Everaert (2021). Risk posed by microplastics: Scientific evidence and public perception. *Current Opinion in Green and Sustainable Chemistry*, 29, 100467.

Chang, Michelle (2015). Reducing microplastics from facial exfoliating cleansers in wastewater through treatment versus consumer product decisions. *Marine Pollution Bulletin*, 101 (1), 330–333.

Chowdhury, Parnali Dhar, C. Emdad Haque, and S. Michelle Driedger (2012). Public versus expert knowledge and perception of climate change-induced heat wave risk: A modified mental model approach. *Journal of Risk Research*, 15 (2), 149–168.

Connor, Melanie, and Michael Siegrist (2011). Factors influencing peoples' acceptance of gene technology: The role of knowledge, health concerns, naturalness, and social trust. *Science Communication*, 32, 514–538.

Cullen, Alison C., C. Leigh Anderson, Pierre Biscaye, and Travis W. Reynolds (2018). Variability in cross-domain risk perception among smallholder farmers in Mali by gender and other demographic and attitudinal characteristics. *Risk Analysis*, 38 (7), 1361–1377.

Davidson, Debra J., and William R. Freudenburg (1996). Gender and environmental risk concerns: A review and analysis of available research. *Environment and Behavior*, 28 (3), 302–339.

De Groot, Judith I., and Linda Steg (2007). Value orientations and environmental beliefs in five countries: Validity of an instrument to measure egoistic, altruistic and biospheric value orientations. *Journal of Cross-Cultural Psychology*, 38 (3), 318–332.

Demski, Christina, Stuart Capstick, Nick Pidgeon, Robert G. Sposato, and Alexa Spence (2017). Experience of extreme weather affects climate change mitigation and adaptation responses. *Climatic Change*, 140 (2), 149–164.

Deng, Lingzhi., Lu Cai, Fengyun Sun, Gen Li., and Yue Che (2020). Public attitudes towards microplastics: Perceptions, behaviors and policy implications. *Resources, Conservation and Recycling*, 163, 105096.

Dilkes-Hoffman, Leela, Steven Pratt, Bronwyn Laycock, Peta Ashworth, and Paul A. Lant (2019a). Public attitudes towards plastics. *Resources, Conservation and Recycling*, 147, 227–235.

Douglas, Mary, and Aaron Wildavsky (1982). *Risk and culture: An essay on the selection of technological and environmental dangers*. Berkeley, CA: University of California Press.

Drummond, Caitlin, and Baruch Fischhoff (2019). Does "putting on your thinking cap" reduce myside bias in evaluation of scientific evidence? *Thinking & Reasoning*, 25 (4), 477–505.

Entman, Robert M., and Andrew Rojecki (1993). Freezing out the public: Elite and media framing of the US anti-nuclear movement. *Political Communication*, 10, 155–173.

Egger, Matthias, Rein Nijhof, Lauren Quiros, Giulia Leone, Sarah-Jeanne Royer, Andrew C. McWhirter, Gennady A. Kantalov, Vladimir I. Radchenko, Evgeny A. Pakhomov, Brian P. V. Hunt, and Laurent Lebreton (2020). A spatially variable scarcity of floating microplastics in the eastern North Pacific Ocean. *Environmental Research Letters*, 15 (11), 114056.

Ericsson, K. Anders, and Mark C. Fox (2011). Thinking aloud is not a form of introspection but a qualitatively different methodology: Reply to Schooler (2011). *Psychological Bulletin*, 137 (2), 351–354.

European Commission (2017). Special Eurobarometer 468: Attitudes of European citizens towards the environment. Brussels. http://ec.europa.eu/comm frontoffice/publicopinion/index.cfm/ResultDoc/download/DocumentKy/12 59.

European Commission (2019). Special Eurobarometer 501, attitudes of European citizens towards the environment. Brussels. 2020. 10.20.2022 https://doi.org/10.2779/902489

Eyal, Tal, Nira Liberman, and Yaacow Trope (2008). Judging near and distant virtue and vice. *Journal of Experimental Social Psychology*, 44 (4), 1204–1209.

Finucane, Melissa L., Ali Alhakami, Paul Slovic, and Stephen M. Johnson (2000). The affect heuristic in judgments of risks and benefits. *Journal of Behavioral Decision Making*, 13 (1), 1–17.

Fischhoff, Baruch, Paul Slovic. and Sarah Lichtenstein (1978). Fault trees: Sensitivity of estimated failure probabilities to problem representation. *Journal of Experimental Psychology: Human Perception and Performance*, 4, 330–344.

Fischhoff, Baruch, Paul Slovic., and Sarah Lichtenstein (1979). Weighing the risks: Risks: Benefits which risks are acceptable?. *Environment: Science and Policy for Sustainable Development*, 21 (4), 17–38.

Frijda, Nico H. (2007). *The laws of emotion*. New York: Psychology Press.

German Federal Institute for Risk Assessment (BfR), Department of Chemicals and Product Safety, Berlin, Germany, Beneventi, E., T. Tietz„ and S. Merkel (2020). Risk assessment of food contact materials. *EFSA Journal*, 18, e181109.

Greven, Frans E., Liesbeth Claassen, Fred Woudenberg, Frans Duijm, and Danielle Timmermans (2018). Where there's smoke, there's fire: Focal points for risk communication. *International Journal of Environmental Health Research*, 28 (3), 240–252.

Grünzner, Maja, Sabine Pahl, Matthew White, and Richard C. Thompson (2021). Experts perceptions about microplastic pollution: potential sources and solutions. [Conference presentation], June 14–16, Finland: Society for Risk Analysis-Europe Conference

Grünzner, Maja, and Sabine Pahl (2022). Behavior change as part of the solution for plastic pollution. In Johanna Kramm and Carolin Völker (eds.). *Living in the plastic age. Perspectives from humanities, social sciences and environmental sciences*. Frankfurt, New York: Campus.

Henderson, Lesley, and Christopher Green (2020). Making sense of microplastics? Public understandings of plastic pollution. *Marine Pollution Bulletin*, 152, 110908.

Heß, Maren, Carolin Völker, Nicole Brennholt, Pia Maria Herrling, , Henner Hollert, Natascha Ivleva, Jutta Kerpen, Christian Laforsch, Martin Löder, Sabrina Schiwy, Markus Schmitz, Stephan Wagner, and Thorsten Hüffer (2022). Microplastics in the Aquatic Environment. In Johanna Kramm and Carolin Völker (eds.). *Living in the plastic age. Perspectives from humanities, social sciences and environmental sciences*. Frankfurt, New York: Campus.

Johnson, Eric J., and Amos Tversky (1983). Affect, generalization, and the perception of risk. *Journal of Personality And Social Psychology*, 45 (1), 20–31.

Kahneman, Daniel, and Amos Tverski (1979). Prospect theory: An analysis of decision under risk. *Econometrica*, 47 (2), 363–391.

Kellstedt, Paul M., Sammy Zahran and Arnold Vedlitz (2008). Personal efficacy, the information environment, and attitudes toward global warming and climate change in the United States. *Risk Analysis*, 28, 113–126.

Kortenkamp, Katherine V., and Collen F. Moore (2010). *Psychology of risk perception*. Hoboken, New Jersey: Wiley Encyclopedia of Operations Research and Management Science.

Kramm, Johanna, and Carolin Völker (2018). Understanding the risks of microplastics: a social-ecological risk perspective. In Martin Wagner and Scott Lambert (eds.). *Freshwater microplastics*, 223–237. Cham: Springer.

Kramm, Johanna, Stefanie Steinhoff, Simon Werschmöller, Beate Völker, and Carolin Völker (2022). Explaining risk perception of microplastics: Results from a representative survey in Germany. *Global Environmental Change*, 73, 102485.

Lerner, Jennifer S., and Dacher Keltner (2001). Fear, anger, and risk. *Journal of Personality and Social Psychology*, 81 (1), 146–159.
Lorenzoni, Irene, and Nick F. Pidgeon (2006). Public views on climate change: European and USA perspectives. *Climatic Change*, 77 (1), 73–95.
Mead, Margaret, and Rhoda Métraux (1954). *Themes in French culture*. Palo Alto, CA: Stanford University Press.
Meijnders, Anneloes L., Cees J. Midden, and Henk A. Wilke (2001). Role of negative emotion in communication about CO2 risks. *Risk Analysis*, 21 (5), 955–955.
Moore, Colleen F. (2009). *Children and pollution: why scientists disagree*. Oxford: Oxford University Press.
Moxnes, Erling, and Carla Susana Assuad (2012). GHG taxes and tradable quotas, experimental evidence of misperceptions and biases. *Environmental Economics*, 3 (2), 44–56.
Nakayachi, Kazuya (2013). The unintended effects of risk-refuting information on anxiety. *Risk Analysis*, 33 (1), 80–91.
Nam, Changhyun, Dong, Huanjiao, and Young A. Lee (2017). Factors influencing consumers' purchase intention of green sportswear. *Fashion and Textiles*, 4 (1), 1–17.
Nardi, Vinicius Antonio Machado, Rafael Teixeira, Wagner Junior Ladeira, and Fernando de Oliveira Santini (2020). A meta-analytic review of food safety risk perception. *Food Control*, 112 107089.
O'Brien, Joshua, and Gladman Thondhlana (2019). Plastic bag use in South Africa: Perceptions, practices and potential intervention strategies. *Waste Management*, 84, 320–328.
Pahl, Sabine, and Kayleigh Wyles (2017). The human dimension: How social and behavioural research methods can help address microplastics in the environment. *Analytical Methods*, 9 (9), 1404–1411.
Petty, Richard E, and John T. Cacioppo (1986). The elaboration likelihood model of persuasion. *Advances in Experimental Social Psychology*, 19, 124–129.
Phelan, Anna (Anya), Helen Ross, Novie Andri Setianto, Kelly Fielding, and Lengga Pradipta (2020). Ocean plastic crisis—Mental models of plastic pollution from remote Indonesian coastal communities. *PLOS ONE*, 15 (7), e0236149.
Poortinga, Wouter, Lorraine Whitmarsh, Linda Steg, Gisela Böhm, and Stephen D. Fisher (2019). Climate change perceptions and their individual-level determinants: A cross-European analysis. *Global Environmental Change*, 55, 25–35.
Renn, Ortwin, and Bernd Rohrmann (eds.). (2000). *Cross-cultural risk perception: A survey of empirical studies*. Dordrecht: Springer Science.
Rist, Sinja, Bethanie Carney Almroth, Nanna B. Hartmann, and Therese M. Karlsson (2018). A critical perspective on early communications concerning human health aspects of microplastics. *Science of the Total Environment*, 626, 720–726.
Rivers, Louie, Joseph Arvai, and Paul Slovic (2010). Beyond a simple case of black and white: Searching for the white male effect in the African-American community. *Risk Analysis*, 30 (1), 65–77.

SAPEA (2019). *A scientific perspective on microplastics in nature and society.* Berlin: SAPEA-Science Advice for Policy by European Academies.

Rohrmann, Bernd, and Huichang Chen (1999). Risk perception in China and Australia: an exploratory crosscultural study. *Journal of Risk Research*, 2 (3), 219–241.

Sabine, Pahl, Bonny Hartley, and Thompson, Richard C. (n.d.). Communicating about marine litter: Insights from the European Marlisco Project. 2. Commocean.org. Available online: https://commocean.org/sites/commocean.org/files/public/docs/abstracts/0712_1530_B2_PahlSabine.pdf.

Satterfield, Terre A., C. K. Mertz, and Paul Slovic (2004). Discrimination, vulnerability, and justice in the face of risk. *Risk Analysis*, 24 (1), 115–129.

Savadori, Lucia, Stefania Savio, Eraldo Nicotra, Rino Rumiati, Melissa Finucane, and Paul Slovic (2004). Expert and public perception of risk from biotechnology. *Risk Analysis*, 24 (5), 1289–1299.

Scott, Sydney E., and Paul Rozin (2020). Actually, natural is neutral. *Nature Human Behavior* 4 (10), 989–990.

Schwartz, Shalom H. (1977). Normative influences on altruism. *Advances in Experimental Social Psychology*, 10, 221–279.

Severin, Marine Isabel, Alexander Hooyberg, Gert Everaert, and Ana Isabel Catarino (2022). Using citizen science to understand plastic pollution: Implications for science and participants. In Johanna Kramm and Carolin Völker (eds.). *Living in the plastic age. Perspectives from humanities, social sciences and environmental sciences.* Frankfurt, New York: Campus.

Shi, Jing, Vivianne H. Visschers, and Michael Siegrist (2015). Public perception of climate change: The importance of knowledge and cultural worldviews. *Risk Analysis*, 35 (12), 2183–2201.

Siegrist, Michael, Philipp Hubner, and Christina Hartmann (2018). Risk prioritization in the food domain using deliberative and survey methods: Differences between experts and laypeople. *Risk Analysis*, 38, 504–524.

Siegrist, Michael, and Joseph Árvai (2020). Risk perception: Reflections on 40 years of research. *Risk Analysis*, 40 (S1), 2191–2206.

Siegrist, Michael, and Christina Hartmann (2020). Consumer acceptance of novel food technologies. *Nature Food*, 1, 343–350.

Sjöberg, Lennart (1998). Risk perception: Experts and the public. *European Psychologist*, 3 (1), 1–12.

Sjöberg, Lennart, and Anders af Wåhlberg (2002). Risk perception and new age beliefs. *Risk Analysis*, 22 (4), 751–764.

Sjöberg, Lennert, and Britt-Marie Drottz-Sjöberg (2001). Fairness, risk and risk tolerance in the siting of a nuclear waste repository. *Journal of Risk Research*, 4 (1), 75–101.

Slovic, Paul (1987). Perception of risk. *Science*, 236 (4799), 280–285.

Slovic, Paul (1999). Trust, emotion, sex, politics, and science: Surveying the risk-assessment battlefield. *Risk Analysis*, 19 (4), 689–701.

Slovic, Paul E. (2000). *The perception of risk*. London, UK: Earthscan publications.
Slovic, Paul, Melissa L. Finucane, Ellen Peters, and Donald G. MacGregor (2004). Risk as analysis and risk as feelings: Some thoughts about affect, reason, risk, and rationality. *Risk Analysis*, 24 (2), 311–322.
Slovic, Paul (2016). Understanding perceived risk: 1978–2015. *Environment: Science and Policy for Sustainable Development*, 58 (1), 25–29.
Shi, Xingmin, Lifan Sun, Xieyang Chen, and Lu Wang (2019). Farmers' perceived efficacy of adaptive behaviors to climate change in the Loess Plateau, China. *Science of the Total Environment*, 697, 134217.
Smith, Nicholas, and Helene Joffe (2013). How the public engages with global warming: A social representations approach. *Public Understanding of Science*, 22 (1), 16–32.
Soares, Joana, Isabeñ Miguel, Cátia Venâncio, Isabel Lopes, and Miguel Oliveira (2021). Public views on plastic pollution: Knowledge, perceived impacts, and pro-environmental behaviours. *Journal of Hazardous Materials*, 412, 125227.
Steentjes, Katharine, Nick F. Pidgeon, Wouter Poortinga, Adam J. Corner, A. Arnold, Gisela Böhm, Claire Mays, Marco Sonnberger, Michael Ruddat, Marc Poumadere, Dirk Scheer and Endre Tvinnereim (2017). *European perceptions of climate change (EPCC): Topline findings of a survey conducted in four European countries in 2016*. Cardiff, UK: Cardiff University.
Steg, Linda, and Judith I. M. de Groot (2018). *Environmental psychology. An introduction*. Second edition. Hoboken, NJ: Wiley-Blackwell; John Wiley & Sons Inc.
Stern, Paul (2000). Toward a coherent theory of environmentally significant behavior. *Journal of Social Issues* 56 (3), 407–424
Tanner, Alexa, and Joseph Arvai (2018). Perceptions of risk and vulnerability following exposure to a major natural disaster: The Calgary flood of 2013. *Risk Analysis*, 38, 548–561.
Teigen, Karl Halvor, Wibecke Brun, and Paul Slovic (1988). Societal risks as seen by a Norwegian public. *Journal of Behavioral Decision Making*, 1 (2), 111–130.
Thacker, Ian, and Gale Sinatra (2019). Visualizing the greenhouse effect: Restructuring mental models of climate change through a guided online simulation. *Education Sciences*, 9 (1), 14.
Thompson, Michael, and Aaron Wildavsky (1982). A proposal to create a cultural theory of risk. In Howard C. Kunreuther and Eryl V. Ley (eds.). *The risk analysis controversy*, 145–161. Heidelberg, Germany: Springer.
Triebskorn, Rita, Thomas Braunbeck, Tamara Grummt, Lisa Hanslik, Sven Huppertsberg, Martin Jekel, Thomas P Knepper, Sstefanie Krais, Yanina Müller, Marco Pittroff, Aki S. Ruhl, Hannah Schmieg, Christoph Schür, Claudia Strobel, Martin Wagner, Nicole Zumbülte and Heinz R. Köhler (2019). Relevance of nano- and microplastics for freshwater ecosystems: A critical review. *TrAC Trends in Analytical Chemistry*, 110, 375–392.

Tsiotsou, Rodoula (2006). The role of perceived product quality and overall satisfaction on purchase intentions. *International Journal of Consumer Studies*, 30 (2), 207–217.

Tversky, Amos, and Daniel Kahneman (1974). Judgment under uncertainty: Heuristics and biases. *Science*, 185 (4157), 1124–1131.

van der Linden, Sander (2015). The social-psychological determinants of climate change risk perceptions: Towards a comprehensive model. *Journal of Environmental Psychology*, 41, 112–124.

Van Sebille, Erik, Chris Wilcox, Laurent Lebreton, Nikolai Maximenko., Britta Denise Hardesty, Jan A. Van Franeker, Marcus Eriksen, David Siegel, Francois Galgani and Kara Lavender Law (2015). A global inventory of small floating plastic debris. *Environmental Research Letters*, 10 (12), 124006.

Völker, Carolin, Johanna Kramm, and Martin Wagner (2020). On the Creation of Risk: Framing of Microplastics Risks in Science and Media. *Global Challenges*, 4 (6), 1900010.

Wyles, Kayleigh J., Sabine Pahl, Matthew Holland, and Richard C. Thompson (2017). Can beach cleans do more than clean-up litter? Comparing beach cleans to other coastal activities. *Environment and Behavior*, 49 (5), 509–535.

Xie, Belinda, Marilyinn B. Brewer, Brett K. Hayes, Rachel I. McDonald, and Ben R. Newell (2019). Predicting climate change risk perception and willingness to act. *Journal of Environmental Psychology*, 65, 101331.

Yoon, Ahyoung, Daeyoung Jeong, and Jinhyung Chon (2021). The impact of the risk perception of ocean microplastics on tourists' pro-environmental behavior intention. *Science of the Total Environment*, 774, 144782.

Everyday Life With Plastics: How to Put Environmental Concern into Practice(s)

Immanuel Stieß, Luca Raschewski, Georg Sunderer

Plastics in Everyday Life Practices

The public perception of plastics is marked by considerable ambivalence. On the one hand, plastics are seen as a superior, flexible, malleable, and highly adaptable material that can be used for an almost infinite number of applications and goods. Being used as packaging material, plastics enable hygienic transport and storage of goods. With the rise of consumer culture, products made of plastics permeated almost all realms of everyday life such as cooking, clothing, personal hygiene, leisure or sports. On the other hand, mass consumption and the flourishing of plastic objects provoked unease and concerns. Plastics are perceived as artificial and synthetic, as opposed to objects crafted by materials taken from nature such as wood, paper, metal or glass. In comparison to these items, plastic objects are often considered to be cheap or of inferior quality. They are perceived as short-lived disposable things (Shove et al. 2007). As environmental awareness grows, plastics are increasingly associated with (unresolved) disposal problems of mass consumption and become emblematic of waste and pollution of the natural environment (Hawkins 2012; Hawkins et al. 2013; 2015).

Given the ever-increasing use of plastics in everyday life, practitioners have to tackle this ambivalence in their own practices. They are called upon to balance their own wants, needs, and expectations with the concerns raised by environmental activists, scientists, and mass media about plastics as a major threat to the global environment. To adjust one's own practices in line with these claims is far from being an easy task. The use of plastics is closely interlinked with the fabric of everyday life. Getting rid of plastics not only requires a replacement of things, but in many cases also entails a change of habits in terms of acquiring, using, maintaining, mending, and disposed of these items.

So far, studies into the perception of environmental risks of plastics have mainly focused on plastic packaging (Heidbreder et al. 2019) or microplastics (Kramm et al. 2022) They show that awareness of the environmental risks associated with plastics is rather high, but only few succeed in changing their own practices correspondingly (Heidbreder et al. 2019; Grünzner and Pahl in this volume). The fact that a change of practices does not even occur when there is a high level of problem awareness can be attributed to a number of reasons: Important barriers are perceived to be practicality and convenience of plastics in the context of everyday life, a lack of knowledge about alternatives to plastic products or a lack of opportunities to implement them, firmly anchored habits of use and disposal, and a shifting of responsibility to other actors (Heidbreder et al. 2019).

Aims and Scope of the Empirical Study

Risk Awareness and Everyday Life Practices

Against this background, we present and discuss the results from an empirical study on how the awareness of environmental risks of microplastics is linked to everyday life practices. In this contribution, we go beyond issues of packaging, investigating the everyday practices of using plastic products that can intentionally or unintentionally lead to a release of microplastics in aquatic ecosystems such as ditches, rivers, ponds or lakes. In this context, we furthermore explore consumers' knowledge and problem awareness about the environmental hazards of microplastics in aquatic environments and their evaluation of these effects. Taking as example selected practices, we investigate the willingness to adopt less environmentally harmful usage and disposal patterns and identified promoting and inhibiting factors that would support or hinder these shifts. Besides taking personal action, consumers might delegate responsibility for the environmental risks of microplastics to other actors. Therefore, we explore how consumers assess their own role in tackling the environmental impacts of microplastics and which other actors they consider to be responsible in this regard. Finally, we discuss information needs and channels of consumers related to microplastics in order to better understand how knowledge on environmental impacts can be provided to foster a shift towards less harmful usage and disposal practices.

The concept of social practices was based on the study of consumption and environmental change (Røpke 2009; Brand 2011; Shove et al. 2012; Sattlegger et al. 2020). Given the plurality of approaches, the concept of social practices is used in different ways. Nevertheless, one common understanding is that a practice is "a routinized behavior in which bodies are moved, objects are handled, subjects are treated, things are described and the world is understood" (Reckwitz 2002, 250). According to Shove et al. (2012), social practices can be defined as routinized and collectively shared patterns of doings that are reproduced and changed by everyday actions of people or "practitioners". In contrast to many other social theories, in practice theory, materiality in the sense of a connection with the physical world plays an important role. Social practices link socio-cultural-symbolic elements (orientations, knowledge, meanings) with material elements (bodies, objects, infrastructures, and natural resources).

Product Groups

The practices under study are linked to four selected product groups. All practices are directly or indirectly related to the entry pathway of microplastics into urban waters.

Peelings

Cosmetic products are frequently cited as the cause of microplastics inputs. Microplastics can be contained in the products as particles or in dissolved form (Bertling et al. 2018b). Cosmetics manufacturers have responded to the ongoing criticism with a voluntary commitment to reduce microplastics in cosmetics. According to this, no more microbeads should be used in cosmetics by 2020. However, microplastics that do not have an exfoliating function, leave-on cosmetic products, and dissolved, gel-like or waxy polymers have so far been exempt from the commitment.

Moist Toilet Paper

Personal hygiene and care wipes are largely made of water- and tear-resistant nonwovens, sometimes incorporating synthetic fibers (made of fossil-based substances such as polymers, polypropylene, polyethylene, polyester). After their use, they are often disposed of in the toilet and not in the household waste as intended (Bundestags-Drucksache 2016). Being part of wastewater, they become a problem especially in wastewater drainage as they can clog and block pumps. Polyester and viscose fibers, for example, have been detected in a blockage of a pump station, which can be traced back to baby wipes (Hübner and Wörner 2016). As a result of using processed synthetic fibers, degradation to microplastics can occur. Since there is little scientific evidence on the extent to which microplastics from wet wipes pose a risk to water bodies and oceans after passing through the sewage treatment plant, it is not possible to conduct a comprehensive risk assessment of wet hygiene and care wipes (Bundestags-Drucksache 2016).

Fleece Textiles

70 percent of all fibers produced worldwide are of biotic or synthetic origin (UBA 2016). A large proportion of clothing contains synthetic fibers. Microfibers are dissolved out with every wash cycle. According to a study by Hartline (2016), the quantity of microfibers released with each wash cycle corresponds to an average of 0.3 percent of the weight of the dry garment. The type of fiber used and the way of washing influence the quantities of microfibers released (Napper and Thompson 2016) The addition of fabric softeners seems to tend to increase the release (Dris et al. 2016). Microplastics from microfibers which can be attributed to the washing of clothes were detected in wastewater treatment plants as well as in marine habitats (Hartline et al. 2016; Dris et al. 2016). Estimates suggest that about two percent of the microplastics released in Germany are due to the abrasion of fibers during washing. The largest share comes from washing clothes in private households (Bertling et al. 2018a).

Dog Poop Bags

The use of dog poop bags is widespread. About 285 million dog poop bags are issued by German cities and municipalities per year, corresponding to about 14 percent of all plastic bags in Germany (Krämer 2020). Dog poop bags are either made of paper, bioplastics (degradable, primarily from starch blends, or biobased, but not degradable) or polyethylene (PE) with a recycled content of 0–100 percent. Although alternatives are available, products made of PE continue to account for over 95 percent of the market volume today. Because of the low material thickness, dog poop bags can be shredded relatively quickly by (mechanical) environmental influences such as wind and waves. If the bags are made of PE, the resulting small parts cannot be biodegraded and contribute to a release of microplastics into the environment. The fate of filled bags in the environment cannot be regarded as negligible, as a study in Hamburg demonstrates (Krämer 2020). Since a large number of poop bags were found in green spaces often located at water bodies, an entry into the limnic system via surface runoff may occur.

To sum up, this contribution investigates i) every day practices related to the above introduced products, ii) explores consumer's knowledge and problem awareness regarding microplastics, and iii) how they assess their own options for action to avoid products releasing microplastics.

Methodology

Sample and Data Collection

The empirical study was conducted as an online survey in February 2020. The sample consisted of the user groups of the four products. Users of a product were defined by the frequency of usage: For peelings, moist toilet paper and fleece jackets must be used frequently or at least occasionally. In the case of poop dog bags, interviewees should walk the dog at least once a week and use poop dog bags. Interviewees were selected by means of a screening questionnaire from a quota sample of the German resident population aged 18 and over. The sample was obtained from an online access panel of Bilendi GmbH. The characteristics of gender, age, school education, size of town and federal state were used for quota sampling. Deviations from the quota sample were corrected by weighting. The field work was conducted by aproxima—Gesellschaft für Markt- und Sozialforschung, Weimar.

The questionnaire comprised different parts: First, all users were asked about their knowledge and attitudes towards microplastics and requirements for an eco-label. A second section which had to be answered by the corresponding users contained specific questions on each of the four product types. Interviewees were asked about only one product, even if they used multiple products.

In total, the sample included of n = 2,000 product users (weighted 2,027). The case numbers for the individual groups are:

– Fleece jackets: 1,456 (1,475)
– Moist toilet paper: 1,095 (1,116)
– Peelings: 909 (949)
– Poop dog bags: 747 (747)

Sample Structure

A striking feature of the sociodemographic structure of the sample is that there is a preponderance of female users. This predominance is particularly pronounced in the case of peelings. The proportion of women here is 72 percent. The age characteristic shows that the users of scrubs and dog excrement bags are younger compared to the other two product groups and the age distribution in the overall population. The age groups 18 to 29 years and 30 to 44 years are both overrepresented, while the group over 60 years is underrepresented. In the other two product categories, however, the average age is higher. Here, the proportion values for the individual age groups are in general relatively close to the values for the overall population. In the case of school education, the proportion of people with a secondary school leaving certificate is somewhat higher in the case of the product categories scrub and dog poop bags, both in comparison with the other two product categories and with the average population. This is understandable in view of the younger age structure for these products, since the proportion of people with a secondary school leaving certificate is generally lower among younger people. Furthermore, it can be seen that for the dog poop bag category, the proportion of households with children is slightly higher compared to the other three product groups.

Consumer Awareness on Microplastics in Water Bodies

Awareness for the Environmental Impacts of Microplastics

At a first glance, there is a broad public awareness about the environmental impact of microplastics. The topic of microplastics in water bodies is also very present. In the survey, almost 90 percent of respondents stated they had already heard something about this topic. A broad majority is convinced that microplastics can cause considerable damage to the environment: Around 90 percent of respondents say that microplastics have a great (45 percent) or very great (47 percent) effect on the environment. Only eight percent rate the effect as minor or non-existent.

Widespread awareness is not necessarily accompanied by comprehensive knowledge of the issues. This becomes apparent when assessing the individual level of information on the subject of microplastics. Only eight percent of respondents said they feel very well informed about microplastics in water bodies. 51 percent feel well informed. 41 percent feel not so well or not at all informed. Interviewees most frequently learned about microplastics in aquatic environments on television, at 90 percent. The second most frequent source of information was the Internet, wherefrom 78 percent said they had learned about this issue. Conversations with friends and acquaintances (63 percent) and conversations with one's own family (58 percent) are other important sources of information. The role played by media varies between the types of media and the age of the respondents. On average, only 40 percent of respondents use social media as their previous sources of information. Among those under 29, however, the figure is 74 percent.

Attitudes Towards Microplastics

Concerns about environmental problems usually encompass different levels or components of engagement. According to the three-component model from classical attitude theory (Rosenberg and Hovland 1960), attitudes can be measured on the basis of cognitive, affective, and conative reactions. On a cognitive level, the main question is whether a problem or threat situation is seen at all. In addition, an affective concern associated with the problem and, in the sense of a conative assessment, the evaluation of the need for action have also to be considered. This is in line with concepts and studies on general and domain-specific environmental awareness, which have a long tradition in environmental behavior research (see Best 2011; Scholl et al. 2016). General environmental awareness refers to attitudes toward overall environmental issues, while domain-specific considerations refer to attitudes toward environmental issues in specific behavioural domains (e.g., mobility or consumption) (Götz et al 2011). Probably the most prominent environmental awareness concept in the German-speaking world is that of Diekmann and Preisendörfer (1998) which is still widely used today.

In accordance with these considerations, attitudes towards microplastics in waters were to be analyzed as a specific environmental concern, including three components: insight into the problem situation, emotional concern and perceived need for action (Tab. 1).

Tab. 1: Cognitive, affective, and conative attitudes on microplastics. (Source: own draft)

Cognitive
I am sure that microplastics in water can become dangerous for human beings at some point.
I am convinced that microplastics in water pose a great danger to our ecosystems.
Affective
I find it worrying that microplastics can end up in our waters.
The issue of microplastics in water scares me a lot.
Conative
Out of responsibility for future generations, we have to do something about microplastics in water today.
Even if the risks posed by microplastics in water are not yet entirely clear, we must take precautionary action.
I am sure that, using advanced technologies, we will also solve the problem of microplastics in water.
Water quality in Germany is very well controlled—I have confidence in that.

Furthermore, the questionnaire also included statements representing barriers to action, such as lack of self-efficacy, overburdening, dilemma of action, or ignorance. The items are largely reformulated statements from an earlier project on micropollutants, in which a similar operationalization was carried out in relation to the problem of pharmaceutical residues in water (see Götz et al. 2013).

Factor analysis was used to examine the dimensional structure of the attitude items. The results indicate that the items can be assigned to three factors (see Tab. 2).

Tab. 2: Factors related to the concern for microplastics in water.

Factor name	Items
Disquiet and fear in the face of the problem	I find it worrying that microplastics enter our waters. The issue of microplastics in water scares me a lot. I am sure that microplastics in water can become dangerous for humans at some point. I am convinced that microplastics in water pose a great danger to our ecosystems. Out of responsibility for future generations, we have to do something about microplastics in water today. Even if the risks posed by microplastics in water are not yet entirely clear, we must take precautionary action.
Confidence in institutions and technology	I am sure by using advanced technology, we will also solve the problem of microplastics in water. Water quality in Germany is very well controlled—I have confidence in that.
Problem denial due to overstraining	It is too much for me to worry about microplastics in water. We consumers can't do anything about the problem of microplastics in water. If others don't join in, I don't see the point of doing anything about microplastics in water. If I don't use products with microplastics and others do, I'm the fool. I would prefer to know nothing at all about such issues.

The first factor "disquiet and fear in the face of the problem" is composed of three content aspects, representing the belief that microplastics in water pose a danger to humans and the environment, an emotional concern of alarm to fear and a precautionary orientation that something must be done about microplastics in water, even if the dangers are not entirely clear now.

The second factor includes items that express relativization based on "confidence in institutions and technology". The implicit message is that there is no urgent need for consumers to contribute to solving the problem. Instead, there is a belief that existing institutions have the problem under control or will solve it in combination with new technologies.

The third factor "problem denial due to overstraining" includes the surveyed barriers to action.

The survey results show that the factor "disquiet and fear in the face of the problem" is relatively strong. For almost all items belonging to this factor, around 50 percent of respondents "fully agree" and another 40 percent "tend to agree". Only the statement "the issue of microplastics in water scares me a lot" deviates from this pattern. Overall, however, two-thirds of respondents agree here as well, which is a remarkable result, since this item addresses a relatively strong emotion.

With regard to the factor "confidence in institutions and technology," it is apparent that there is a relatively strong tendency in this direction. In each case, just under 20 percent fully agree and around 50 percent tend to agree.

The barriers to action belonging to the factor "problem denial due to overstraining" are seen by 20 to 25 percent of respondents in each case (fully agree and tend to agree). Only for the item "I would prefer to know nothing at all about such issues" there is somewhat less agreement (15 percent).

The expression of the factors differs between sociodemographic groups: On average, women have a slightly higher concern for the environmental threats of microplastics in water (factor 1) than men. They are less likely to put the problem into perspective by trusting institutions and technology, and show to a lesser extent a denial of the problem. Among the age groups, the 30- to 45-year-olds have the lowest concern of the problem and most frequently show a problem denial (factor three). One possible reason for this could be that the double load of work and family care that often exists in this life span poses a burden, making other problems fade into the background. Among the age groups, the older cohorts have the highest concern and the lowest level of denial, while the youngest cohorts (18–29 years) are in the middle.

Overall, the survey shows a fairly pronounced concern for the risks of microplastics in the general public. As we learned from many studies, a high environmental concern is important, but only one prerequisite for behavior change. Changing practices is a much more complex process (see also Grünzner and Pahl in this volume). In the following section, we will explore closer whether and how the concern for microplastics could lead to changes in consumption practices.

Microplastics in Consumption Practices

Given the markedly high level of awareness of the problem of microplastics in waters, the question arises as to how products that are made of plastics are dealt with in everyday life. Consumption practices can contribute to the release of microplastics into the environment in different ways. The input can occur directly through products that contain microplastics. Or microplastics are created in connection with the use of products or through the degradation of plastics after their disposal. Depending on how the release and entry occurs, different options for action arise. In the following part of our contribution we will examine practices of acquiring, using, mending, and disposing of products that are made of plastics or contain microplastics. The main objective is to explore to what extent consumers are prepared to make changes in their consumption practices in order to prevent environmental damage by microplastics.

Peelings

Peelings are used by a considerable part of consumers. 35 percent of persons in the screening sample use peelings at least occasionally. The frequency of application varies. Half of the users surveyed use peelings once or several times a week, another third once to three times a month, and the rest less frequently. Peeling is usually applied in the shower or at the sink, removed mostly with the help of water, sometimes also with cosmetic tissues.

The majority of users knows the fact that peelings can contain microplastics. Almost 60 percent have already heard of it. To a certain extent, this concern is also reflected in the relevant criteria for product selection. Decisive criteria when buying peelings are health aspects: 70 percent rate the skin compatibility of the peeling as "very important". In second place are the criteria ingredients and "free of microplastics" (both 46 percent "very important"). But consumers show also a striking degree of uncertainty as to whether their own peeling meets these requirements. 37 percent of respondents are certain that their peelings does not contain microplastics, while 56 percent say that they do not know this for sure.

To prevent microplastics enter into water through peelings, an obviously simple and apparent solution is not to use products that contain microplastics. A majority of 62 percent of users says that this option would be certainly suitable for them. In this regard making a peeling themselves is considered rather attractive. 32 percent mentioned that this is definitely a suitable option for them. Reducing the usage of peelings, appears to be less attractive: Using peelings less frequently is only a definite option for 17 percent and only seven percent would dispense with peelings altogether.

In addition to individual action, respondents were asked to rate selected measures of other actors that would prevent the release of microplastics in waters from peelings. 35 percent assessed a governmental ban of adding microplastics to products as the most important measure. For 24 percent a voluntary waiver by manufacturers is key. A claim on the product's packaging whether the peeling contains microplastics or not is the most preferred measure for 18 percent. Technical solutions, such as the installation of an additional filter stage in wastewater treatment plants, have the highest preference for 14 percent.

Furthermore, respondents assessed to what extent they see different societal actors responsible for reducing the release of microplastics into the water cycle through peelings. 74 percent consider manufacturers of peelings as having a very strong responsibility. 48 percent attribute a strong responsibility to policy makers and 46 percent to themselves, i.e. each individual consumer. Only 19 percent see a strong responsibility on the part of water and wastewater treatment plant operators.

Moist Toilet Paper

The use of moist toilet paper has increased considerably in recent years. 41 percent of persons in the screening sample say they use this product at least occasionally. When disposing of moist toilet paper, pumps and agitators can become clogged. Only few users are aware that moist toilet paper is made of plastic fibers, which can be crushed into microplastics in the sewage system. 66 percent of users of moist toilet paper have not heard of this problem. 73 percent of respondents assume or tend to assume that moist toilet paper dissolves in the sewage system. 48 percent consider the problem to be (rather) negligible. Interest in finding out more about the problem is correspondingly low. Only a quarter say they want to find out more about microplastics from moist toilet paper.

The main requirement for moist toilet paper is to be skin-friendly, which is very important to 62 percent of respondents. Other important criteria are tear resistance (39 percent) and a favorable price (37 percent). Environmental friendliness is seen as less significant (30 percent). 31 percent find it very important that the product should be biodegradable.

Moist toilet paper is mainly disposed of like conventional toilet paper. 58 percent of users say that they always throw it into the toilet. Only a smaller proportion of users (38 percent) states that they put the wipes at least occasionally in the residual waste. Many respondents would stick to their disposal habits: 25 percent surveyed would be prepared to consistently dispose of moist toilet paper in the residual waste. 44 percent of users cannot imagine to change their habits. To those who are accustomed to moist toilet paper, the use of dry toilet paper appears hardly as a viable alternative. Only 19 percent surveyed definitely agree with this solution, 43 percent rather agree.

Biodegradable moist toilet paper could be an option to prevent the release of microplastics. The survey looked more closely at what consumers know about this option and how they evaluate it. As it turned out, only few users have a clear idea of what biodegradability is about. Only 38 percent of respondents say they know what the term biodegradable means. 42 percent had already heard about it, but had only a rough idea and 19 had no idea or had not yet heard about the term at all.

Considering what action other stakeholders should take to prevent microplastics from entering water, respondents prefer strategies that eliminate plastic fibers from moist toilet paper. 31 percent of users rate a governmental ban on the use of plastics in moist toilet papers as most important. 29 percent opt for a voluntary waiver by industry 21 percent advocate better labelling on the packaging, with information about the ingredients and instructions for an appropriate disposal through residual waste.

When asked whom consumers consider to be most responsible for preventing microplastics from entering waterways, manufacturers are named most frequently. 69 percent see manufactures as most responsible, followed by policymakers and consumers with 47 percent. Sewage treatment plants bring up the rear. 17 percent rate them as most responsible to reduce the entry of microplastics in water through moist toilet paper.

Fleece Jackets

Fleece clothing is widely used in Germany. According to the screening sample, 56 percent of the population own one or more garments made of fleece. Fleece clothing is used in many ways. 79 percent use a fleece jacket as a cozy leisure jacket, 61 percent for hikes and walks, just over half wear a fleece jacket when doing housework or gardening, and 21 percent at work.

Decisive for the choice of fleece jackets are aspects such as fit, quality, material, functionality, and design. But price, or a good price-performance ratio, is also central. These are followed by the aspects of eco-friendliness (for 70 percent "very important/important") and fair production conditions (67percent). The country of manufacture also has a certain importance for 46 percent.

Fleece jackets are laundered rather frequently, with 36 percent washing them at least every two weeks and another third every two to four weeks. 43 percent say they air or brush it out when it is only lightly soiled. Laundry is mostly done at low temperatures 51 percent of respondents use temperatures up to 30 degrees, another third wash at temperatures between 30 and 40 degrees. 81 percent dry fleece garments in the air and only six percent consistently use a tumbler.

Fleece jackets and sweaters are considered as cozy and robust clothing. That they release microplastics when washed is only little known: 58 percent of those who use fleece textiles had not heard about this issue before. 22 percent firmly believe that their fleece jacket is an eco-friendly product (recycled plastic). For many, the issue of microplastics loss is considered to be of little concern: 46 percent view the problem to be rather negligible and only 22 percent are sure they want to find out more about it.

The users of fleece clothing were asked to rate possible actions they could take to prevent microplastics from entering the water. The most popular options with the highest approval were: Air dry fleece jackets rather than try them in the tumbler (64 percent), do not empty the lint filter of the washer and tumbler down the drain (57 percent), and wash fleece jackets only when really necessary (55 percent). More consistent solutions that would significantly minimize the input of microplastics are only viable for a small group of users. 26 percent of the respondents can definitely imagine using a wash bag, 33 percent would buy fleece jackets made of natural fibers, and only 22 percent would buy fleece jackets from manufacturers who offer products that release less microplastic.

26 percent consider the most important action to be an obligation on the part of manufacturers to provide product labelling informing that fleece jackets can release microplastics. A legal ban on plastic fibers in fleece jackets, which release microplastics, triggers strongly polarizing reactions: 25 percent rate a ban to be most important, while 37 percent considers this to be the least important action. 20 percent consider the installation of an additional filter stage that can remove microplastics from wastewater to be particularly important.

Considering the responsibility of societal actors, the respondents see it above all to be the duty of manufacturers to work on the problem. 62 percent say that manufacturers have a strong obligation to reduce the release of microplastics from fleece textiles. 45 percent consider it to be their responsibility, for example, to adapt their washing habits or to pay attention to this when buying. 41 percent of respondents attribute a strong obligation to policymakers, and 24 percent to the operators of sewage plants.

Dog Poop Bags

The use of dog poop bags is widespread. According to the screening sample, they are used by 27 percent of the population in Germany. 75 percent of the dog owners surveyed said they carry a dog poop bag with them when they walk their dogs. But only half of them is employing the bag every time. For the users of dog poop bags, material durability is the most important property: 89 percent consider it important or very important that dog poop bags are reliably tear-resistant and not too thin. 80 percent of users state that it is very important or important to them that the bags are made of biodegradable material.

A large majority of dog owners feel responsible to remove the dog's excrement in order to keep the landscape clean. They comply with this as a matter of course. However, 78 percent complain that there are too few trash cans for easy and uncomplicated disposal. In many cases, they find it very inconvenient to have to carry the full bag around for a long time. A small minority of 10 percent admit to sometimes illegally dropping dog poop bags in the landscape. 26 percent say they use a bag mainly when they might be caught leaving the droppings lying around.

The majority of users is aware that dog poop bags can become a source of microplastics: 62 percent have already heard of microplastics and the dog poop bag problem, and 92 percent consider this issue to be very important or important. Easily accessible dispenser facilities are named as the most important measures against uncontrolled disposal. A large majority of 95 percent is in favor of placing more trash cans in places where they use to walk their dog. What is also frequently mentioned is the use of dog poop bags made of biodegradable material. 65 percent of users believe this to be a reasonable solution. 15 percent already use biodegradable bags and another 65 percent state they would certainly do so. 59 percent of users consider it very important that only dog waste bags made of biodegradable plastics are offered. However, not all of them would accept additional costs for such a bag. Only 28 percent of respondents state they would certainly be willing to pay more for biodegradable products than for those made from conventional plastics, 43 percent say they rather would do.

Considering who should act to prevent microplastics entering the environment through dog poop bags, 57 percent of respondents see themselves, i.e. the individual consumers, as being strongly responsible. Manufacturers of dog poop bags are also seen as having a strong responsibility with 56 percent. 47 percent attribute a strong obligation to legislators and policymakers, and 40 percent to municipalities as providers of the infrastructure for dispensing and disposing of dog poop bags. The lowest degree of responsibility is assigned to water and sewage treatment plants. Only 24 percent of respondents consider them to have a very strong responsibility.

Lessons Learnt from the Analysis of Practices

The four examples show a significant discrepancy between consumers' overall concern with microplastics and their own handling of products that can release microplastics. Despite the high level of problem awareness, consumers assess their own options for action very differently—depending on the product group. One obvious explanation is that in many cases consumers are not familiar with the links between their own practices and the release of microplastics in the environment. While a majority of users of peelings and dog poop bags are aware of this connection, many users of fleece jackets and moist toilet papers have not heard about it. Many respondents were not aware that moist toilet paper contain plastic fibers at all. A lack of knowledge about a product's properties and its impacts on the environment appears to be an important barrier to change.

But the investigation also shows that many consumers are prepared to change their routines when dealing with plastics in products. Examples include avoiding cosmetics that contain microplastics, shifting practices for cleaning synthetic fiber textiles (airing instead of washing), or reducing littering by properly disposing of dog waste bags. A change in usage and disposal practices seems most likely to be acceptable if, as in the case of peelings, a direct contribution of one's own behavior to the input of microplastics is recognizable and viable alternatives are at hand. In the case of dog poop bags, a reduction of littering or the usage of biodegradable bags play a similar role. Apparently, a shift towards more sustainable practices is dependent on how available services support or constrain them: Many users said that more and better available dispensers and disposal facilities would help reduce littering of dog poop bags. In other cases, consumers have only restricted options for action because—as in the case of moist toilet paper—the ingredients of products are not known or existing alternatives, such as the use of wash bags, are difficult to reconcile with daily routines. In addition, supposedly environmentally friendly options, such as the use of products made of biodegradable plastic, might themselves have problematic environmental impacts.

For this reason, it seems important to take a broader perspective. Public discourse on microplastics mainly evolved around the narrative of consumers as polluters. The responsibility for solving the microplastics problem is essentially attributed to consumers, and is to be achieved through behavioral changes (Beyerl et al. 2022). Less attention is paid to other actors such as manufacturers, disposal companies or regulators. These actors are important stakeholders, because they have the power to design how plastics is built in products and how it is to be disposed of. They have a significant influence on how entries into the environment can be prevented. The survey clearly shows whom consumers consider to be responsible for solving the microplastics problem: For most of the products studied, manufacturers were attributed the greatest responsibility for preventing microplastics discharges into waters. Policymakers and consumers themselves are named as responsible somewhat less frequently and at an almost equal level. Operators of wastewater treatment plants are far less often seen as responsible. The only exception is the example of dog waste bags, where a majority sees consumers themselves to have a strong obligation to act compared to all other stakeholders.

From the consumer's point of view, there is a need for cooperation among different stakeholder groups to reduce the release of microplastics into water. Consumers are willing to contribute their part to solving the microplastics problem, but they expect transparency and environmentally friendly products from industry. Regulations by the legislator, especially towards the manufacturers, are also considered to be important. "End-of-pipe" solutions, such as water treatment technologies, are only seen by a minority to be a way out of this the problem.

Eco-labelling as a Tool for Consumer Information on Microplastics?

Given the apparent uncertainties among many consumers as to whether products contain or release microplastics, a label could be an option to provide product-related information on microplastics. Eco-labels and quality seals are important instruments for conveying knowledge about environmentally relevant properties and impacts of products and services in a simple and comprehensible form. They can make complex information on environmental impacts easier to grasp in everyday life situations. However, given the large number of already existing eco-labels and seals issued from a multiplicity of bodies and organizations, it is not easy to find a path through the "seal jungle".

Against this background, it is an open question whether an eco-label on microplastics might be helpful to support consumers who want to choose products and adopt practices that reduce the release of microplastics in water. To this end, all respondents of the survey were asked how important they thought an eco-label guaranteeing that a product does not release microplastics was. Overall, the idea of such a label was viewed quite positively. 53 percent of respondents consider such an eco-label to be very important and a 36 percent rate it as important.

The second question was how such a label could be designed. Two options were presented for this purpose: Firstly, a new, stand-alone label, and secondly, an extension of an existing eco-label, e.g., the Blue Angel in Germany, with an additional claim on microplastics. All respondents who consider an eco-label to be less important, important or very important were asked what type of label they would prefer. Overall, there is tendency towards a stand-alone level but the picture is not so clear: 43 percent favored a stand-alone eco-label claiming that the product does not release microplastics. 22 percent of respondents preferred the extension of an existing eco-label. 32 percent find both options equally appealing.

When asked who could issue such an eco-label, a clear majority is in favor of the federal government. 66 percent see the German Federal Environment Ministry and the Federal Environment Agency as a trustful and reliable agency of such a label. 44 percent see environmental associations (e.g., WWF or Greenpeace) as suitable. Manufacturers and industry or consumer organizations are named by around 40 percent of respondents.

These findings show a strong desire for better and easier orientation and a decision-making aid. Consumers would mostly appreciate a specific label on microplastics, but the extension of an already existing label would also find acceptance.

Given the plethora of available products that are made of plastics or of plastics components, it appears not easy to define suitable criteria that would match all relevant product properties. Available labels on microplastics are rather doubtful or even misleading. For example, the claim "free of microplastics" only considers primary microplastics and does not provide any information about of how the product degrades in an aquatic environment. A simple claim of "free from microplastics" seem appealing, but would not include the release of microplastics during the use of a product or when it is disposed of.

In sum, a label indicating that a product is free from microplastics would hardly meet the complex information needs. Considering the rather vague understanding of biodegradability among consumers, a better indication is needed of how products made from biodegradable plastics behave under specific environmental conditions, in particular in aquatic environments. Furthermore, additional information is required on specific product properties such as the durability of fibers in textiles or on the absence of environmentally harmful additives in plastic products. This information would be useful for consumers, but also for other stakeholder, like retailers or manufacturers who process plastic materials in their products (Plastrat 2021).

An eco-label on microplastics should be only one element of a more comprehensive communication. In order to encourage consumers to be more aware of the use of plastic products and materials, target-group-specific communication on the problem of microplastics in water is needed. Communication should highlight the areas in which consumers have the opportunity to reduce the input of microplastics into water significantly. Recommendations should focus on alternatives products that do not contain or release microplastics or that are made of comparatively safe materials. In addition to information about product properties alternative practices for caring, mending, and disposal of products would help change everyday life practices.

Author Contribution

Immanuel Stieß conceptualized the outline and wrote the final manuscript. Luca Raschewski contributed an earlier draft of the manuscript and provided empirical data. Georg Sunderer supervised the empirical fieldwork of the survey and provided the statistical analysis and interpretation of data.

Acknowledgements

Our acknowledgement goes to Konrad Götz and Barbara Birzle-Harder from the PLASTRAT research team for their valuable and inspiring advice and support of the empirical fieldwork.

The project *PLASTRAT: Lösungsstrategien zur Verminderung von Einträgen von urbanem Plastik in limnische Systeme, Teilprojekt 3 Societal perception and assessment systems* has received funding from the German Federal Ministry on Education and Research (BMBF) under the grant 02 WPL1446 C. Responsibility for the information and views set out in this contribution lies entirely with the authors.

References

Bertling, Jürgen, Leandra Hamann, and Ralf Bertling (2018a). *Kunststoffe in der Umwelt*. Oberhausen: Fraunhofer Umsicht.

Bertling, Jürgen, Leandra Hamann, and Markus Hiebel (2018b). *Mikroplastik und synthetische Polymere in Kosmetikprodukten sowie Wasch-, Putz- und Reinigungsmitteln*. Oberhausen: Fraunhofer Umsicht.

Best, Henning (2011). Methodische Herausforderungen – Umweltbewusstsein, Feldexperimente und die Analyse umweltbezogener Entscheidungen. In Matthias Groß (ed.). *Handbuch Umweltsoziologie*, 240–258. Wiesbaden: VS Verlag.

Beyerl, Katharina, Franz Bogner, Maria Daskalakis, Thomas Decker, Anja Hentschel, Mandy Hinzmann, Bastian Loges, Doris Knoblauch, Linda Mederake, Ruth Müller, Frieder Rubik, Stefan Schweiger, and Immanuel Stieß (2022). Wege zum nachhaltigen Umgang mit Kunststoffen: Kernbotschaften sozialwissenschaftlicher Forschung. *GAIA,* 31 (1), 51–53.

Brand, Karl-Werner (2011). Umweltsoziologie und der praxistheoretische Zugang. In Matthias Groß (ed.) *Handbuch Umweltsoziologie,* 173–198. Wiesbaden: VS Verlag.

Bundestags-Drucksache 18/10761 (22. Dezember 2016): Antwort der Bundesregierung auf die Kleine Anfrage der Abgeordneten Birgit Menz, Ralph Lenkert, Caren Lay, Eva Bulling-Schröter und der Fraktion. 17.07.2022 http://dipbt.bundestag.de/dip21/btd/18/107/1810761.pdf.

Diekmann, Andreas, and Peter Preisendöfer (1998). Umweltbewußtsein und Umweltverhalten in Low- und High-Cost-Situationen: Eine empirische Überprüfung der Low-Cost-Hypothese. *Zeitschrift für Soziologie*, 27 (6), 438–456.

Dris, Rachid, Johnny Gasperi, Mohamed Saad, Cécile Mirande, and Bruno Tassin (2016). Synthetic fibers in atmospheric fallout: A source of microplastics in the environment? *Marine Pollution Bulletin,* 104, 290–293.

Götz, Konrad, Jutta Deffner, and Immanuel Stieß (2011). Lebensstilansätze in der angewandten Sozialforschung, am Beispiel der transdisziplinären Nachhaltigkeitsforschung. *Kölner Zeitschrift für Soziologie und Sozialpsychologie,* Special Issue 51/2011, 86–112.

Götz, Konrad, and Barbara Birzle-Harder (2013). Spurenstoffe im Wasserkreislauf - Wahrnehmung, Reaktion, Handeln. Presentation held at a TransRisk project meeting. March 6–7, Darmstadt, Germany.

Niko L. Hartline, Nicholas J. Bruce, Stephanie N. Karba, Elizabeth O. Ruff, Shreya U. Sonar, and Patricia A. Holden (2016). Microfiber masses recovered from conventional machine washing of new or aged garments. *Environmental Science and Technology,* 50, 11532–11538.

Grünzner, Maja, and Sabine Pahl (2022). Behavior change as part of the solution for plastic pollution. In Johanna Kramm and Carolin Völker (eds.). *Living in the plastic age. Perspectives from humanities, social sciences and environmental sciences.* Frankfurt, New York: Campus.

Hawkins, Gay (2012). The performativity of food packaging. Market devices, waste crisis and recycling. *The Sociological Review* 60, 66–83.

Hawkins, Gay, Emily Potter, and Kane Race (2015). *Plastic Water: The Social and Material Life of Bottled Water Opens in a New Window.* Massachusetts: The MIT Press, Cambridge.

Hawkins, Gay, Jennifer Gabrys, and Mike Michael: (2013). *Accumulation: The material politics of plastic.* London: Routledge.

Heidbreder, Lea Marie, Isabella Bablok, Stefan Drews, and Claudia Menzel (2019). Tackling the plastic problem: A review on perceptions, behaviors, and interventions. *Science of The Total Environment*, 668, 1077–1093.

Hübner, Mario, and Horst Wörner (2016). Abwasserpumpen mögen keine Feuchttücher! *Verfahrenstechnik, Special Pumpen*, 7–8/2016, 8–14.

Krämer, Andrea (2020). *Hundekotbeutel – Hintergrund & Problematik*. https://thesustainablepeople.com/wp-content/uploads/Hundekotbeutel-Hintergrund-und-Problematik-FrpercentC3percentBChjahr-2020.pdf.

Napper, Imogen E., and Richard C. Thompson (2016). Release of synthetic microplastic fibres from domestic washing machines: Effects of fabric type and washing conditions. *Marine Pollution Bulletin*, 112, 39–45.

Reckwitz, Andreas (2002). Toward a theory of social practices. *European Journal of Social Theory* 5 (2), 243–63.

Røpke, Inge (2009). Theories of practice. New inspiration for ecological economic studies on consumption. *Ecological Economics*, 68 (10), 2490–2497.

Rosenberg, Milton J., and Carl I. Hovland (1960). Cognitive, affective and behavioral components of attitudes. In Milton J. Rosenberg, and Carl I. Hovland (eds.). *Attitude organization and change. An analysis of consistency among attitude components*. New Haven, CT: Yale University Press.

Sattlegger, Lukas, Immanuel Stieß, Luca Raschewski, and Katharina Reindl (2020). Plastic packaging, food supply, and everyday life. Adopting a social practice perspective in social-ecological research. *Nature and Culture*, 15 (2), 146–172.

Schaum, Christian, Steffen Krause, Natalie Wick, Jörg Oelmann, Ulrike Schulte-Oelmann, Kristina Klein, Immanuel Stiess, Luca Raschweski, Georg Sunderer, Barbara Birzle-Harder, Kristina Wencki, Peter Levai, Hans-Joachim Maelzer, Gerhard Schertzinger, Helena Pannekens, Elke Dopp, Thomas Ternes, Georg Dierkes, Peter Schweyen, and Juliana Ivar do Sul (2021). *Lösungsstrategien zur Verminderung von Einträgen von urbanem Plastik in limnische Systeme – PLASTRAT–Synthesebericht*. Mitteilungen – Institut für Wasserwesen. Band 134. Neubiberg: Universität der Bundeswehr. München.

Scholl, Gerd, Maike Gossen, Brigitte Holzhauer, und Michael Schipperges (2016). *Mit welchen Kenngrößen kann Umweltbewusstsein heute erfasst werden? – Eine Machbarkeitsstudie*. Texte 58/2016, Dessau-Roßlau: Umweltbundesamt.

Shove, Elisabeth, Matt Watson, Martin Hand, and Jack Ingram (2007). The Materials of Material Culture: Plastic. In Elisabeth Shove, Matthew Watson, Martin Hand, and Jack Ingram (eds.). *The design of everyday life*. 94–116. Oxford: Berg Publishers.

Shove, Elizabeth, Mika Pantzar, and Matt Watson (2012). *The dynamics of social practice. Everyday life and how it changes*. Los Angeles: SAGE.

UBA (2016). *Mikroplastik: Entwicklung eines Umweltbewertungskonzepts. Erste Überlegungen zur Relevanz von synthetischen Polymeren in der Umwelt*. Texte 32 / 2016. Dessau-Roßlau: Umweltbundesamt.

Using Citizen Science to Understand Plastic Pollution: Implications for Science and Participants

Marine Isabel Severin, Alexander Hooyberg, Gert Everaert, Ana Isabel Catarino

Using Citizen Science to Understand Plastic Pollution

Plastic pollution is ubiquitous in aquatic environments around the world, such as in lakes, rivers, and oceans, and the presence and accumulation of plastic litter in coastal environments has become a topic of high priority for policymakers. The proliferation of plastic litter is mostly driven by an increasing production of synthetic-based polymers combined with poor waste management strategies (Moore 2008). An estimate of 8 million metric tonnes of macroplastic (plastic debris > 5 mm) and 1.5 metric tonnes of microplastic (plastic debris < 5 mm) enter the oceans annually (Jambeck et al. 2015; Boucher and Friot 2017), and these values have been projected to double by 2050 under a business-as-usual scenario (Lau et al. 2020). Plastic pollution can be detrimental to marine ecosystems due to the potential hazardous effects on biota and ecosystem functions, potentially affecting life history of (commercial) fish species and degrading habitats. The estimate of environmental damage to marine ecosystems is 13 billion US dollars per year, including financial losses incurred by fisheries and tourism as well as time spent cleaning up beaches (UNEP 2014). Beaumont et al. (2019) estimated that each metric tonne of plastic discarded in the marine environment has an economic cost of 3,300−33,000 US dollars per year. To assess the current extent of plastic litter in the environment, surveys and monitoring programs have been implemented, but remain challenging due to the geographical extent and persistent nature of plastic pollution, which leads to accumulation of litter items, combined with often limited financial and time resources (GESAMP 2019). Citizen science has, however, the potential to address data gap related-issues, by using participants as "sensors" and simultaneously increasing the public awareness towards plastic pollution.

Citizen science projects related to marine sciences are popular (e.g., current estimate of 500 marine coastal projects actively running in Europe, Garcia-Soto et al. (2021)), including projects in the field of plastic pollution, which have recently increased in number (Rambonnet et al. 2019). For example, of 127 citizen science projects active in the North Sea in 2020, 17 percent focused on marine pollution, including plastic debris observations (van Hee et al. 2020). According to current guidelines and best practice suggested by the European Citizen Science Association (ECSA), in citizen science projects participants are included in one or more stages of the research process, and projects should have scientific outcomes, which are genuinely open access (ECSA 2015; Heigl et al. 2019; Haklay et al. 2021). Citizen science initiatives, related to the observation and mitigation of plastic pollution, actively involve the public in scientific research and in the policy-making process (Lippiatt et al. 2013; GESAMP 2019; Garcia-Soto et al. 2021). The benefits of citizen science are multifold. Firstly, data acquired by citizen scientists can fill in gaps in existing data and information about plastic pollution levels due to resource constraints (time, staff, etc.), and enables sampling over larger geographical areas (Rambonnet et al. 2019). Secondly, the data acquired by citizen science participants, obtained either by schoolchildren and/or adults, is often of equivalent quality to that collected by experts (Falk-Andersson et al. 2019; van der Velde et al. 2017). Consequently, such initiatives contribute significantly to obtaining information on plastic debris distribution (GESAMP 2019). For example, data acquired by citizen science can be instrumental in rapid assessment surveys, i.e., in obtaining an initial "snap-shot" of the distribution and abundance of marine litter (GESAMP 2019). Finally, plastic pollution data obtained by volunteers can also be further useful to evaluate the effectiveness of mitigation actions and local environmental policies, such as recycling initiatives (Harris et al. 2021; Lippiatt et al. 2013). However, while the benefits for scientific outputs have been well described, the impact for the participants in citizen science projects are less known.

Citizen science projects have per definition a strong component of public engagement in the scientific process, and some studies suggest that participation can lead to increased awareness of environmental issues, such as plastic pollution. For instance, participation in citizen science projects has a positive impact on the public's scientific literacy in several ways. Volunteers report a positive shift in their attitude towards science and gain an increased understanding of the nature of science. Participants also obtain topic-specific knowledge of the project and the steps of the scientific method (Bonney et al. 2009; Cronje et al. 2011; Aristeidou and Herodotou 2020; Peter et al. 2021). Further reasons and motivation to participate include enhancement of career competencies, environmental concerns, and interest in science (West et al. 2021). Recent reports further indicate that volunteers working outdoors experience a sense of enjoyment and of satisfaction when participating in citizen science projects as well as an increased connection to people and nature (Peter et al. 2021). Regarding activities specifically related to plastic pollution, Locritani et al. (2019) demonstrated that school students change their perception on beach debris causes, sources, transport, and consequences after participating in citizen science projects. For young adults (e.g., university students), participation in beach clean-up activities have been associated with positive mood, pro-environmental intentions, and higher marine awareness (Wyles et al. 2017).

The impact of public participation in citizen science related to plastic pollution on health and well-being is largely unknown. We define the term health as the overall absence of illness, injury, or pain and well-being as "a state of happiness and contentment, with low levels of distress, overall good physical and mental health and outlook, or good quality of life" (American Psychological Association 2021, n.p.). To date, there is little knowledge on how public beach clean-ups and plastic surveying activities promote human health and well-being as well as the participants' ocean literacy compared to other recreational visits to the coast. From an Ocean and Human Health perspective, a meta-discipline that explores the link between the health of the ocean and that of humans, citizen science projects have an important role in promoting ocean literacy. Besides, they offer an excellent opportunity to explore the benefits and risks for the participants' health and well-being after their interaction with the ocean and other water features (blue spaces) (H2020 SOPHIE Consortium 2020).

The goal of this chapter is to review the current knowledge on the impact of citizen science activities related to plastic pollution (e.g., beach clean-ups, plastic surveying, etc.) on participants, specifically science literacy, awareness of plastic pollution, public health, and well-being. We further discuss the benefits of the citizens' participation in plastic pollution related projects from an Ocean and Human Health perspective.

Methodology

Systematic Literature Review

We performed a systematic literature search on 19th May 2021, following the Preferred Reporting Items for Systematic Reviews and Meta-Analyses (PRISMA) 2020 statement (Page et al. 2021, prisma-statement.org), using the abstract and citation database Scopus (www.scopus.com; Elsevier) (Tab. 1, Appendix A). The goal of the literature search was to retrieve peer-reviewed publications on citizen science activities, related to plastic pollution (Search 1), with an additional discussion on scientific literacy and attitudes and behaviors towards marine litter (Search 2) (Tab. 1). The queries used Boolean operators, and the selected terms were searched in the field "title, abstract, and keywords" (TITLE-ABS-KEY). Both searches were merged, and we obtained a result of 56 publications (Search 1, n = 37; Search 2, n = 19) (Fig. 1). After the exclusion of duplicates (n = 9), the remaining 47 publications were then manually screened. Publications that were further excluded (n = 13) from the results included those not focusing on citizen science activities (e.g., were about other educational interventions), reviews, and other non-research peer-reviewed articles (e.g., methods development). We obtained a final list containing 34 peer-reviewed research publications (Fig. 1).

Non-Systematic Literature Search

To assess whether citizen science studies further measured the impact of the activities on the participants, we performed an additional non-systematic literature free-text search using Google Scholar (freely accessible metadata of scholarly literature, scholar.google.com, Google) and Scopus. The non-systematic literature search retrieved four peer-reviewed articles (Tab. 1).

Data Visualization

All data was visualized using the package ggplot2 (Wickham 2016) from R (R Core Team 2020).

Tab. 1: Number of peer-reviewed articles retrieved in the systematic literature search (Search 1 and 2) and non-systematic search. Queries of Searches 1 and 2 used Boolean operators and terms were searched in the "title, abstract, and keywords" (TITLE-ABS-KEY) in Scopus (Elsevier, 19/05/2021), whereas the non-systematic search was done using free-text terms in Google Scholar and Scopus. See Appendix A for the flow of the Preferred Reporting Items for Systematic Reviews and Meta-Analyses (PRISMA).

Search	No. of Results	Plastic Pollution related terms		Citizen Science terms		Ocean Literacy/ Education related terms
Search 1	37	(TITLE-ABS-KEY ("plastic pollution") OR TITLE-ABS-KEY ("plastic litter") OR TITLE-ABS-KEY ("plastic debris") OR TITLE-ABS-KEY ("plastic cleanup") OR TITLE-ABS-KEY ("beach cleanup"))	+	TITLE-ABS-KEY ("citizen science")		
Search 2	19	TITLE-ABS-KEY ("plastic pollution") OR TITLE-ABS-KEY ("litter") OR TITLE-ABS-KEY ("beach cleanup")	+	TITLE-ABS-KEY ("citizen science")	+	TITLE-ABS-KEY ("awareness") OR TITLE-ABS-KEY ("education")
Non-systematic search	4	Free-text search(es)				

Results

Systematic Literature Review

The literature search retrieved a total of 34 studies (Appendix) published between 2013 and 2021 (Fig. 1) which targeted citizen science activities mostly in Europe (n = 16), e.g., in Norway (n = 5) and Denmark (n = 4), and North America (n = 9), e.g., USA (n = 5) (Fig. 2). Nine out of the 34 publications reported activities that took place in two or more countries. All studies used citizens as sensors or for surveying/monitoring purposes. In 31 of the 34 retrieved articles, citizens quantified coastal, marine or riverine litter items in a systematic way, following standard operating procedures (SOPs), specified or developed in each study (Tab. 2). These included beach cleanups (n = 8), boat sampling (n = 1), river shore sampling (n = 1), coastal shore sampling (n = 19), using social media records (n = 1), or thanks to an opportunistic and haphazard activity (n = 1) (Tab. 2). In 29 out of the 34 articles, quality control of the sampling procedure and/or established scientific standards for plastic sampling were discussed. Of the 34 articles, only nine specifically stated that the intervention targeted school children and adolescents (minors), whereas most of the retrieved articles did not specify the age of the citizen scientists (and we assumed that they were either adults or a mixed population of all ages) (Tab. 2).

In our results, only two of the retrieved studies assessed the impact of citizen science activities (in short term) on the participants using systematic methodologies. Of these two, one study acquired data via feedback from volunteers and case studies, in the form of online communication and interviews (Yeo et al. 2015). The work of Yeo et al. (2015) took place in Australia and New Zealand, and consisted of surveying persistent organic pollutants (POPs) using plastic resin pellets. The assessment of the citizen scientists' participation in the activity was done by collecting information via email and (online) feedback from the volunteers. The main conclusion of the study by Yeo et al. (2015) was that the participants reported feeling "empowered/encouraged", "grateful", and "more aware about the issue". The authors further concluded that "active participation in citizen science increases the participants' awareness of marine debris issues" (Yeo et al. 2015, 142, 144). The second impact assessment, a study that took place in the Italian coast, used questionnaires to acquire data from students as citizen scientists (Locritani et al. 2019). In their study, Locritani et al. (2019) performed a quantitative assessment of students' attitudes and behaviors towards marine litter before and after their participation in an educational and citizen science project for surveying macro- and micro-litter. The results of the study by Locritani et al. (2019, 320) demonstrated that the students (n = 87) "changed quantitatively their perception of beach-litter causes and derived problems", and that the students "improved their knowledge about the main marine litter sources and the role of the sea in the waste transport and deposition along the coast".

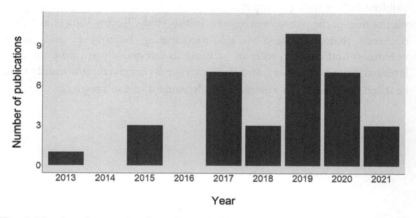

Fig. 1: *Number of peer-reviewed articles per publication year on studies of citizen science activities related to plastic pollution retrieved using a systematic literature search (total n = 34 publications; search performed on 19/05/2021 using Scopus, Elsevier).*

Non-Systematic Literature Search

In the non-systematic literature review, we retrieved four additional peer-reviewed articles published between 2017 and 2020 (Tab. 2), which assessed citizens' perception, attitudes towards, and awareness of coastal litter in both children and adults. Of these four studies, two gave out questionnaires to beachgoers (coastal visitors) (Lucrezi and Digun-Aweto 2020; Rayon-Viña et al. 2018) to quantify the perception, awareness, and behaviors regarding coastal litter of participants undertaking citizen science activities (Rayon-Viña et al. 2019; Lucrezi and Digun-Aweto 2020). Wyles et al. (2017) assessed in university students (n = 90) from the UK, via pre- and post-intervention questionnaires, their marine awareness, behavioral intentions, mood, well-being, and perceived restorativeness of the coast. In their study,

Wyles et al. (2017) observed that participants in beach cleaning activities reported a higher positive mood and pro-environmental intentions after the activity, with no significant differences to individuals who did activities such as rock pooling or coastal walks. Beach cleaning activities were further associated with higher marine awareness and were rated as most meaningful, but linked to lower restorativeness ratings of the environment (Wyles et al. 2017). Rayon-Viña et al. (2019) interviewed Asturian (Spain) adults and children before their participation in a coastal debris sampling campaign as well as non-participating beachgoers. In this study, Rayon-Viña et al. (2019) observed that, compared to non-participating beachgoers and adults, volunteers and children were more likely to erroneously attribute the main litter origin to beachgoers, and that volunteers perceived significantly more beached litter than non-volunteers, independent of the age group.

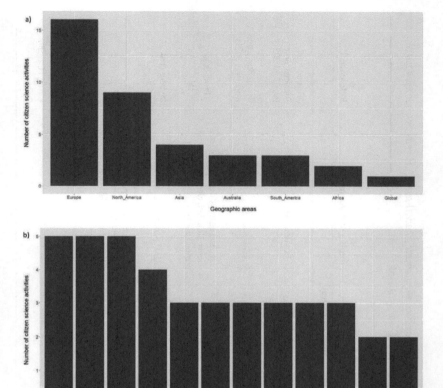

Fig. 2: Number of citizen science activities (absolute frequency) which took place a) around the world ("Global" included two or more geographic areas), and b) per country (only frequencies equal or above two were plotted), from studies retrieved using a systematic review approach [for details on systematic literature search see Tab. 1 and main text].

Tab. 2. *Retrieved peer-reviewed articles, from a systematic (unshaded) and an unsystematic (shaded) search. Results were retrieved via Scopus (Elsevier) on 19/05/2021. SOP, standard operating procedure; NA, not applicable*

Method for litter collection/observation	Target geographic landform	Age group of participants	acquisition (surveys/ questionnaires)	Number of participants	Reference
Search in social media mentions	NA	Not specified	NA	279 respondents	(Turner, Williams, and Pitchford 2021)
Cleanup	Beach	Not specified	NA	NA (meta-analysis)	(Earn, Bucci, and Rochman 2021)
Cleanup	Beach	Not specified	NA	NA (data analysis from eight citizen science shoreline cleanup organizations)	(Harris et al. 2021)
Following a sampling SOP	Coastal shore	School students (6 - 19 y/o)	NA	57,000 participants	(Syberg et al. 2020)
Following a sampling SOP	Coastal shore	Not specified	NA	Over three million person-days (a single individual doing an activity for any portion of a day) of recreational activity	(Uhrin et al. 2020)
Following a sampling SOP	River shore	Not specified	NA	Not specified	(Bernardini, McConville, and Castillo Castillo 2020)

Method for litter collection/observation	Target geographic landform	Age group of participants	Information acquisition (surveys/questionnaires)	Number of participants	Reference
Cleanup	Beach	Not specified	NA	Not specified	(Roman et al. 2020)
Cleanup	Beach	Not specified	NA	Not specified	(Nelms et al. 2020)
Following a sampling SOP	Coastal shore	Not specified	NA	6,102 volunteers	(Chen et al. 2020)
Following a sampling SOP	Coastal shore	Not specified	NA	744 citizen scientists	(Tunnell et al. 2020)
Cleanup	Beach	Not specified	NA	2,139 volunteers	(Mayoma et al. 2019)
Following a sampling SOP	Coastal shore	Not specified	NA	37 surveyors	(Lee, Hong and Lee 2019)
Cleanup	Beach	5–80 y/o	NA	85–183 average daily participants	(Cowger, Gray, and Schultz 2019)
Following a sampling SOP	Coastal shore	Not specified	NA	417 volunteers	(Ambrose et al. 2019)
Cleanup	Beach	Not specified	NA	Not specified	(Turrell 2019)

Method for litter collection/observation	Target geographic landform	Age group of participants	Information acquisition (surveys/ questionnaires)	Number of participants	Reference
Following a sampling SOP	Coastal shore	Not specified	NA	17 groups of citizen scientists	(Forrest et al. 2019)
Following a sampling SOP	Coastal shore	School children	NA	5,500 students	(Kiessling et al. 2019)
Following a sampling SOP	Coastal shore	Not specified	NA	Not specified	(Falk-Andersson et al. 2019)
Following a sampling SOP	Coastal shore	School children (nine–17 y/o)	NA	1,383 students (supported by 80 teachers)	(Honorato-Zimmer et al. 2019)
Cleanup	Beach	Not specified	NA	498 volunteers	(Walther et al. 2018)
Following a sampling SOP	Coastal shore	Not specified	NA	214 participants	(Loizidou et al. 2018)
Opportunistic and haphazard	NA	Not specified	NA	Not specified	(Smith et al. 2018)
Following a sampling SOP (collection by manta trawls attached to standup paddleboards)	Baltic Sea	Not specified	NA	Two "adventurers"	(Gewert et al. 2017)

Method for litter collection/observation	Target geographic landform	Age group of participants	Information acquisition (surveys/ questionnaires)	Number of participants	Reference
Following a sampling SOP	Coastal shore	Not specified	NA	195 participants, including guests, crew and staff of cruises	(Bergmann et al. 2017)
Following a sampling SOP	Coastal shore	Not specified	NA	Not specified	(Lots et al. 2017)
Opportunistic and haphazard	NA	Not specified	NA	Not specified	(Colmenero et al. 2017)
Following a sampling SOP	Coastal shore	Not specified	NA	2,000–6,000 volunteers per year (estimate)	(Nelms et al. 2017)
Following a sampling SOP	Coastal shore	Not specified	NA	600+ volunteers	(Davis and Murphy 2015)
Following a sampling SOP	Coastal shore	School students and adults	Interviews and informal feedback	110 individuals or organizations	(Yeo et al. 2015)
Following a sampling SOP	Coastal shore	School students (eight–16 y/o)	Informal survey	983 students (supported by 43 teachers)	(Hidalgo-Ruz and Thiel 2013)

Method for litter collection/observation	Target geographic landform	Age group of participants	Information acquisition (surveys/questionnaires)	Number of participants	Reference
Following a sampling SOP	Coastal shore	School students (16–17 y/o)	Questionnaires: quant. assessment of attitude and behaviors twd marine litter before/after participation	194 students (87 of which replied further to the post-survey questionnaire)	(Locritani, et al. 2019)
Following a sampling SOP	Coastal shore	School students (primary & secondary) and adults	NA	Not specified	(van der Velde et al. 2017)
Following a sampling SOP	Coastal shore	School students with an average of twelve y/o (other target groups: local resident children and adults)	Assessed perception and knowledge of local communities: 2x questionnaires and follow-up interviews	48 and 60 interviews per region	(Kiessling et al. 2017)
Following a sampling SOP	Coastal shore	School students/children	NA	Not specified	(Mioni et al. 2015)

Method for litter collection/observation	Target geographic landform	Age group of participants	Information acquisition (surveys/ questionnaires)	Number of participants	Reference
NA	NA	Mostly young adults	Quantitative, descriptive and non-experimental research design, structured questionnaire	512 beach visitors	(Lucrezi and Digun-Aweto 2020)
Following a sampling SOP	Coastal shore	Adults and children	Interviews before beach cleanups and survey	75 volunteers (and 133 beachgoers)	(Rayon-Viña et al. 2019)
Following a sampling SOP	Coastal shore	Adults	Survey about perception and awareness (open answers)	201 beachgoers	(Rayon-Viña et al. 2018)
Following a sampling SOP	Coastal shore	Undergraduate university students	Online surveys and paper surveys before activity, other measurements post activity	90 participants	(Wyles et al. 2017)

Discussion

Scientific Relevance of Citizen Science Activities Related to Plastic Pollution

Citizen science activities related to plastic pollution have recently become popular and represent a global effort of engaging the public in collaborative scientific projects. In our review, we demonstrate that citizen science projects related to plastic pollution have led to scientific outputs, i.e., peer-reviewed scientific publications (ECSA 2015). Our data shows that a quarter (nine out of 38) of the assessed studies represented a collaboration between two or more countries, indicating an effort of covering larger geographical areas of sampling. The data on plastic litter acquired by the public are considered of high quality and of scientific value, and comparable to that obtained by trained professionals (e.g., Falk-Andersson et al. 2019). The promotion of data quality and interoperability standards in citizen science depends largely on the development of clear sampling methodologies, with transparency of data management and sharing principles, and methodologies that include quality control and assurance measures (European Commission 2020). Plastic data acquired via citizen science projects can be made available (open data) in non-commercial portals such as The European Marine Observation and Data Network (EMODnet) Chemistry [www.emodnet-chemistry.eu], LitterBase [litterbase.awi.de, (Tekman et al. 2021)] or via mobile apps, such as the Marine Debris Tracker [debristracker.org/data, National Oceanic and Atmospheric Administration, NOAA, USA]. Most of the reported projects in this chapter (76 percent) included a discussion on the quality control and assurance of the obtained data, and followed a systematic sampling protocol for plastic litter quantification (Tab. 2). The quality of surveyed data is key to produce datasets that have comparability and should be considered in the evaluation of such projects in terms of cost-benefits (GESAMP 2019). Even though training citizen scientists can be time-consuming and costly (Hidalgo-Ruz and Thiel 2015; GESAMP 2019), the public participation can cover large geographic areas (e.g., Colmenero et al. 2017; Lots et al. 2017; Turrell 2019; Roman et al. 2020) that would be difficult or expensive for technical staff to visit (Hidalgo-Ruz and Thiel 2015), leading to clear benefits for valuable scientific outputs.

Most of the reviewed studies in this chapter focused on activities that assessed larger marine plastic debris, macroplastics (> 5 mm), and only three studies have quantified microplastics (< 5 mm), a more technically challenging work. The surveying and collection of macroplastics requires easy-to-understand sampling methodologies that do not demand complex equipment nor extensive training (GESAMP 2019). As per our results, citizen scientists successfully engage in projects that use systematic plastic data collection and employ standardized protocols. The collection of microplastic data by citizen scientists requires however, that two important analytical challenges are overcome (Zettler et al. 2017), i.e., inclusion of additional quality control measures (e.g., incorporation of procedural blank samples to quantify background contamination) and avoidance of unintentional microplastics contamination (e.g., airborne fibers) (Forrest et al. 2019). In their work, Lots et al. (2017) combined the collection of beach sediment samples done by citizen scientists with the laboratory extraction of microplastics done by technically trained staff and in a controlled environment. Forrest et al. (2019) used a similar approach to sample microplastics from Ottawa River (Canada). In their work, volunteers filtered a volume of 100 liter of river water using a previously provided sampling kit and then had the samples processed in the laboratory (Forrest et al. 2019). In the case of Gewert et al. (2017), two Swedish "adventurers" provided extra sampling opportunities by collecting microplastics during a paddleboard expedition over the Baltic Sea using lightweight trawls and processing the samples in a laboratory setting. In these studies, the participation of the public enabled researchers to cover a broader geographical area for microplastics sampling: data was collected in 23 locations of 13 countries in Lots et al. (2017), in locations covering a 550 km river stretch in Forrest et al. (2019), and along a transect of 210 km in the Baltic Sea in Gewert et al. (2017). After performing the quality check, four drawbacks of citizen science microplastics sampling compared to researchers taking the samples were mentioned in the discussion sections. These drawbacks mostly related to volunteers not following the step-by-step instructions such as not recording metadata, not processing blank samples, not sampling the required number of replicates or by overestimating the volume of sampled water (Forrest et al. 2019). Even so, for both macro- and microplastics sampling, authors noted that their outputs were reliable and data replicable (e.g., Kiessling et al. 2019; Bernardini et al. 2020), and the collected data provided important indications of litter density and composition (e.g., Bernardini et al. 2020; Harris et al. 2021) to inform recommendations for local and international plastic waste and litter management (Hidalgo-Ruz and Thiel 2015; Bernardini et al.2020; Harris et al. 2021). A multistep data verification flowchart with multiple criteria to be reached as presented in Kiessling et al. (2019) is an excellent example on how to ensure the quality of the citizen science data, and which can assist future projects to report on their data acquisition.

A frequent claim of citizen science litter surveying projects and beach clean-ups is that the public engagement also leads to an increased awareness of plastic pollution, with potential impact on the volunteers' ocean literacy and pro-environmental behaviors (Zettler et al. 2017; GESAMP 2019). Increasing ocean literacy, i.e., the understanding of human impact on the oceans and of the ocean's impact on humans, has been considered to contribute to tackling waste management issues and to improve the understanding of local communities on the potential consequences of plastic (Westfall and Simantel 2019). Even though considerable work is being done on engaging volunteers in scientific projects, there is still a considerable knowledge gap on how plastic related projects impact citizen scientists. For example, in our work, of the 38 publications evaluated (systematic and non-systematic searches), only four studies have assessed the impact of the activity on the participants. The retrieved studies reported, for instance, that participants felt "more aware about the (plastic pollution) issue" (Yeo et al. 2015, 142), and "changed quantitatively their perception of beach-litter causes and derived problems" (Locritani et al. 2019, 320). However, the assessed aspects were not only restricted to the educational benefits and pro-environmental intentions of the activity. For instance, Yeo et al. (2015) and Wyles et al. (2017) further assessed aspects related to mood and well-being of the participants. As with other citizen science projects, plastic related activities are intrinsically transdisciplinary, and an active inclusion of social sciences and humanities questions and methodologies would benefit the understanding of the benefits and challenges of the interventions in the public (Tauginienė et al. 2020). In the following sections, we will discuss the educational and behavioral impact of citizen science as well as the potential psychological implications of the public participation in plastic pollution related projects from an Ocean and Human Health perspective.

The Educational and Behavioral Impact of Citizen Science

Citizen science has important benefits in the development of scientific literacy, beyond the facilitation of data collection or data analysis to reach a specific research outcome. The term scientific literacy can refer to either knowledge of scientific processes, concepts, or situations (OECD 2006) or to awareness of one's role in the local environment (Conrad and Hilchey 2011). In the case of knowledge and awareness of ocean and human interactions, the more common term of "ocean literacy" is used; however, ocean literacy and scientific literacy are seen as interdependent as "one cannot be considered 'science literate' without being 'ocean literate'" (Strang et al. 2007, 7). Enhancing ocean literacy encompasses several processes, such as educating on marine environmental issues, increasing awareness and sensitivity, and developing a connection as well as pro-environmental attitudes and behaviors towards the ocean (Kelly et al. 2022). A study by Ashley et al. (2019) demonstrated that ocean literacy initiatives led to an increased awareness, knowledge, and attitudes, supporting sustainable actions to marine environmental issues that were considered key predictors of behavior change, which is the goal of ocean literacy.

Considering today's context of marine plastic pollution, it would be beneficial to evaluate the capabilities of citizen science interventions to improve knowledge, awareness, and attitudes essential to addressing the problem. Citizen science is already regarded as being of educational value to the participants, however increasing knowledge or awareness is insufficient to predicting actual behavior change (Hines et al. 1987). Changes in attitudes and behavioral intentions are necessary additional predictors of behavior change (Ashley et al. 2019, see also Grünzner and Pahl in this volume), but are often overlooked during evaluation of citizen science interventions (Toomey and Domroese 2013). As per our results, only four studies from the (systematic and non-systematic) literature search assessed the impact of citizen science activity on the participants, although there is an increased effort of the scientific community in recent years to assess the public perception of plastic pollution related issues (Catarino et al. 2021). More investigation on the educational and behavioral effects of citizen science is therefore necessary, to establish the impact of the citizen's participation in sampling campaigns in tackling the plastic pollution problem.

To assess the educational and behavioral impact of citizen science, researchers follow social science methodologies (e.g., questionnaires, interviews, etc.), that can be applied in studies related to plastic pollution. One method that has been implemented by mostly social scientists in numerous studies is a quantitative pre- and post-assessment via a questionnaire (Hartley et al. 2015; 2018; Locritani et al. 2019; Wyles et al. 2017). This type of within-subject design allows to concisely assess the short-term impact of the activity on the participants and thereby evaluating its effectiveness. For example, specific questions regarding knowledge, awareness, attitudes, and behaviors towards marine litter are developed and typically piloted before being used on the participants. Participants then complete the questionnaire before and after the citizen science activity (Breakwell et al. 2006). Another methodology, applied by Rayon-Viña et al. (2019), compared participants taking part in beach clean-ups with non-participants and evaluated potential differences in perception and awareness of marine litter by means of a survey. This between-subject design resembles that of a typical interventional design employed in social sciences with one group undergoing the "intervention" and the other remaining as "control group", enabling to tease out the unique effect of the activity (Breakwell et al. 2006). Notably, the intervention group (i.e., citizen science participants) may differ from the control group in many ways (e.g., demographic, psychological), which may confound the envisioned effects, despite efforts to ensure equal variability in both groups via randomization Another example is the methodology employed by Yeo et al. (2015), i.e., the use of qualitative methods to investigate the impact of the science communication of their program on the volunteers. The researchers analyzed feedback from the volunteers that was sent via email, conducted on-site interviews, and implemented case studies including participant observation (Yeo et al. 2015). This type of methodology provides an open access to the participants' thoughts and feelings concerning the citizen science activity. Although applying qualitative methods does not enable to test a specific effect, it does enable to indicate potentially undiscovered effects that could be interesting to pursue and further test with quantitative methods (Taylor 2005).

According to the results of the presented literature research, there are indications that participation in citizen science activities can affect volunteers. In the study conducted by Wyles et al. (2017), marine awareness and pro-environmental behavioral intention increased significantly after the citizen science activity and remained higher than baseline after one week, suggesting a short-term effect. Additionally, the study's results imply a positive spillover effect, as participants not only expressed a stronger intention to engage in beach clean-ups, but also reported a higher intention to adopt more general pro-environmental behaviors (Wyles et al. 2017). This supports the notion that engaging in one conservation act can induce individuals to undertake other pro-environmental behaviors (Grønhøj and Thøgersen 2012). The effectiveness of citizen science also seems to vary by age group. Essentially using the same questionnaire as Hartley et al. (2015) for ages between eight to 13 years, Locritani et al. (2019) state differences in the effects of their citizen science activity on the adolescent participants, aged 16 to 17 years. The results from Locritani et al. (2019) indicate that adolescents had a higher baseline knowledge and awareness of marine litter pollution, thereby leading to less significant changes between pre- and post-activity. Alongside these results, children aged ten to 16 years had a higher awareness of the problem of marine litter compared to adults in another study, even though this was not linked with a higher litter perception (Rayon-Viña et al. 2019). This suggests that although increasing children's awareness is a priority, there remains a lack of knowledge about marine litter. Although participation in citizen science related activities might affect specific age groups differently, the educational value of engaging in such activities should not be discarded. For example, by directly empowering educators to teach about marine litter, as suggested by Hartley et al. (2018), and by including adults in citizen science projects, knowledge transfer to children can be indirectly promoted.

Health Impacts of Coastal Citizen Science Activities

By participating in citizen science projects, the public can experience benefits to their health and well-being. Two studies in our systematic review reported changes in the health or well-being of the participants (Wyles et al. 2017; Yeo et al. 2015), a key aspect in establishing the link between ocean health and human health. The impact of plastic surveying and beach clean-ups has only been marginally investigated compared to other and more common activities in coastal areas, such as walking or other leisure coastal activities (Maguire et al. 2011; White et al. 2013; Wyles et al. 2014). More specifically, health and well-being aspects that were investigated in these two studies retrieved from the current literature review can be categorized as having an emotional or cognitive origin. The studies reported emotional changes in the participants, including feeling more "empowered/ encouraged" and "grateful" (Yeo et al. 2015, 142), as well as a better mood and higher meaningfulness (Wyles et al. 2017). However, negative aspects were also reported, such as a lower cognitive restoration in response to beach cleaning compared to walking or rock pooling (Wyles et al. 2017), indicating that beach cleaning may affect the participants self-perception of their relaxation status.

The citizen science participants can experience key benefits or shortcomings that can affect their emotions and cognitive restoration from actively participating in plastic surveying and beach clean-up activities. For example, the blue gym concept explains the mechanisms behind the health benefits potentially experienced by citizen science participants' exposure to blue spaces (water features, including coastal areas). The blue gym refers to the use of coastal environments that promotes health and well-being, thanks to physical activity, stress reduction, and by building a community spirit (Sea Change 2018). Participants in plastic pollution surveying and beach clean-up activities are exposed to an outdoors setting (e.g., natural coastal landscape, clean air), participate in a physical activity (e.g., via plastic collection, walking for surveying purposes), and can experience positive social interactions (with friends, family, and other community members). Research from the fields of environmental health and psychology suggests that spending time in a natural setting, such as a beach or coastal environment, can result in better mood (Peng and Yamashita 2016), less stress (Triguero-Mas et al. 2017), and decreased depressive symptoms (Dempsey et al. 2018). However, the relative absence of cognitive restoration from citizen science activities compared to other beach activities found by Wyles et al. (2017) can be attributed to the exposure and focus on the plastic litter. For example, litter seems to negatively impact the perceived restorative potential of a landscape when shown via pictures (Wyles et al. 2016; Hooyberg et al. in prep). Wyles et al. (2016) additionally showed that the litter decreased the preference for the shown environment. Recent research indicates that this effect is especially relevant in natural landscapes compared to urban landscapes and is higher than the impact of other anthropogenic disturbances such as cars (Hooyberg et al. in prep). The fact that citizen science activities related to plastic pollution imply exposing the public to litter may limit the restorative characteristics of the activity. Future research should verify and elucidate these effects in detail.

In contrast with the anticipated positive effects, the participants of citizen science projects have at some occasions reported pessimism, anxiety, and other negative emotions. For example, in an online survey given to participants across 63 biodiversity citizen science projects, the majority of which focused on insects and birds, a small portion of these participants reported a pessimistic outlook regarding the future of the environment after participation (Peter et al. 2021). This is in line with research displaying the negative consequences of environmental education such as pessimism and eco-anxiety, referring to the "chronic fear of environmental doom" (Sheppard 2004; Clayton et al. 2017; Pihkala 2020). As such, there is a small but existing risk for participants in citizen science to develop negative emotions. To make it even more complex, pessimistic emotions and views towards the future can be constructive, in the sense that the more negatively one perceives a future situation, the more likely one will be encouraged to act preventively in the present, as demonstrated by Kaida and Kaida (2016). Further research on the potential negative outcomes of participating in citizen science should be conducted, as addressing these issues could improve the positive outcomes of citizen science projects (Peter et al. 2021).

The repeated exposure to plastic litter during beach clean-ups and other plastic pollution related activities, whether being it in the context of citizen science or not, may induce further psychosocial impacts. Such impacts may arise from coastal users or citizen science participants perceiving that other coastal visitors are responsible for littering, which may in turn impact their social relationships (Wyles et al. 2014). For instance, coastal users have identified activities such as walking to be detrimental to the environment, as they perceived that other coastal visitors are responsible for littering (Wyles et al. 2014). As the perceived restoration of the landscape moderates the effect on pro-environmental behaviors (Berto and Barbiero 2017), a similar outcome can be expected for people participating in citizen science activities related to plastic. The same impact on citizen scientists' perception can occur in the interactions between health-related effects (e.g., restoration) and literacy-related effects. Currently, the health and well-being benefits of citizen science plastic-related activities are poorly quantified, but recent reports indicate that complex psychological, social, and physical factors can jointly and interactively play a role in determining the health benefits of the citizen science plastic activity.

Socio-Economic and Socio-Demographic Representation

There is an inherent bias in the socio-economic and socio-demographic representation of the participants of citizen science activities (Haklay 2013), questioning whether societal and environmental benefits are evenly distributed (Cooper et al. 2021; Pateman et al. 2021). Most participants are from middle to high income backgrounds, with access to education, technical skills, resources, and infrastructure that facilitate engagement in citizen science projects (Haklay 2013; Cooper et al. 2021; Pateman et al. 2021). There are further concerns on how diverse citizen science participants are (Cooper et al. 2021; Pateman et al. 2021). For example, in the UK, participation in environmental citizen science projects is particularly low among women from minority groups and people who are unemployed or from lower socio-economic groups (Pateman et al. 2021). At the same time, the participating population in citizen science activities in the US does not reflect the demographics of this country, mostly excluding individuals from groups that have been historically underrepresented in science (e.g., African Americans, Latinx, American Indigenous Communities) (Pandya 2012; Trumbull et al. 2000). It has been demonstrated, however, that more social deprived communities in the UK have the highest health benefits when living in close proximity to the sea (i.e., blue spaces), potentially thanks to increased opportunities for stress reduction and physical activity (Wheeler et al. 2012). It is critical to further understand whether the participation of lower socio-economic groups in citizen science activities related to plastic pollution would provide additional health and educational benefits to such communities, to promote projects that account for diversity, equity, and inclusion dimensions.

The current literature review highlights that most of the retrieved publications have reported on activities that took place in Europe and North America (Fig. 2), where countries are mostly classified as middle to high-income economies (The World Bank 2021), with Africa as being the continent with the lowest number of reported projects (Fig. 2). We were unable to assess the socio-economic background of the participants in the literature reviewed, as such data are rarely recorded in citizen science outputs (Pandya and Dibner 2018; Pateman et al. 2021). The lack of participation of people from specific geographic areas or from specific communities may imply that certain areas are not considered in environmental datasets and are excluded from prioritization in policies (Pandya and Dibner 2018; Pateman et al. 2021), for example, to prevent and mitigate plastic pollution. Furthermore, members from these communities will not have access to the benefits in terms of skills and literacy gained by participating in citizen science projects, such as having direct contact with scientists or exposure to scientific literacy (Pandya and Dibner 2018; Pateman et al. 2021). Projects such as the Citizen

Observation of Local Litter in Coastal ECosysTems (COLLECT) (Partnership for Observation of the Global Ocean 2021) and the WIOMSA (Western Indian Ocean Marine Science Association) Marine Litter Monitoring Project (Western Indian Ocean Marine Science Association 2021), among others, are important contributors in working towards data acquisition on plastic debris distribution and abundance on the coasts of African countries, by training citizen scientists and promoting knowledge transfer between local communities and researchers. Additionally, COLLECT further aims to evaluate shifts in the pro-environmental attitude of participants towards coastal plastic litter, as well as on their well-being, filling an important knowledge gap on the health and educational impacts of citizen science initiatives in Africa. Overcoming underrepresentation of entire geographic areas and/or of underprivileged and minority groups in plastic pollution related citizen science should be considered in projects, as it is a missed opportunity to empower local communities in taking part in the development of a successful plastic circular economy, via perception and behavior shifts, and via active participation in decision-making.

Conclusions and Outlook

Citizen science projects related to plastic pollution, such as beach clean-ups and litter surveying, have increased in popularity and have produced valuable scientific outputs. The participants in citizen science projects can follow standardized methodologies for data acquisition, and allow to cover a wider geographical range for sampling than conventional observational and monitoring efforts. Even though the scientific advantages of citizen science plastic related projects are well documented, only a limited number of studies have reported on the educational and behavioral effects on participants. There are however indications of the positive educational value for the public in participating in beach clean-ups and plastic surveying activities as well as positive impacts in terms of ocean literacy, pro-environmental behaviors, higher meaningfulness, and general well-being. Some negative impacts have also been suggested, such as induced changes in the participants' emotions, which limit restoration of their cognitive abilities. An important finding of our systematic literature review is that both positive and negative impacts on participants of citizen science plastic projects have barely been explored, and that there is a gap in the knowledge of whether the public experiences pessimism and eco-anxiety feelings. We further identified a limited socio-economic and socio-demographic representation of the participants of citizen science activities in plastic pollution projects. Plastic related projects involving citizen scientists should broaden their collaborative scopes to include geographic areas overlooked in current projects, consider the inclusion and empowerment of diverse groups of participants (and beneficiaries), to deepen the projects' impact, and to avoid important data gaps due to the exclusion of participants due their socio-economic and -demographic status. Future surveying programs of plastic pollution involving citizen scientists should consider a collaboration between natural and social science professionals, to evaluate in depth the educational and psychological benefits to the participants, and that best practices should include mechanisms to engage across diverse publics, including access to activities by underrepresented socio-economic and socio-demographic groups.

CRediT author statement

Marine Isabel Severin: Conceptualization, Investigation, Writing - Original Draft, Writing - Review & Editing; Alexander Hooyberg: Conceptualization, Investigation, Writing - Original Draft, Writing - Review & Editing; Gert Everaert: Conceptualization, Supervision, Writing - Review & Editing; Ana Isabel Catarino: Conceptualization, Investigation, Methodology, Supervision, Writing - Original Draft, Writing - Review & Editing

Marine Isabel Severin and Alexander Hooyberg share first authorship, Gert Everaert and Ana Isabel Catarino share senior authorship. Ana Isabel Catarino is the corresponding author *(ana.catarino@vliz.be)*.

References

Ambrose, Kristal K., Carolynn Box, James Boxall, Annabelle Brooks, Marcus Eriksen, Joan Fabres, Georgios Fylakis, and Tony R. Walker (2019). Spatial trends and drivers of marine debris accumulation on shorelines in South

Eleuthera, The Bahamas Using Citizen Science. *Marine Pollution Bulletin,* 142, 145–54.

American Psychological Association (2021). American Psychological Association dictionary of psychology. 15.07.2021 https://dictionary.apa.org/.

Aristeidou, Maria, and Christothea Herodotou (2020). Online citizen science: A systematic review of effects on learning and scientific literacy. *Citizen Science: Theory and Practice,* 5 (1), 1–12.

Ashley, Matthew, Sabine Pahl, Gillian Glegg, and Stephen Fletcher (2019). A change of mind: Applying social and behavioral research methods to the assessment of the effectiveness of ocean literacy initiatives. *Frontiers in Marine Science,* 6, 288.

Beaumont, Nicola J., Margrethe Aanesen, Melanie C. Austen, Tobias Börger, James R. Clark, Matthew Cole, Tara Hooper, Penelope K. Lindeque, Christine Pascoe, and Kayleigh J. Wyles (2019). Global ecological, social and economic impacts of marine plastic. *Marine Pollution Bulletin,* 142, 189–95.

Bergmann, Melanie, Birgit Lutz, Mine B. Tekman, and Lars Gutow (2017). Citizen scientists reveal: Marine litter pollutes Arctic beaches and affects wild life. *Marine Pollution Bulletin,* 125 (1–2), 535–40.

Bernardini, Giulia, A. J. McConville, and Arturo Castillo Castillo (2020). Macroplastic pollution in the tidal Thames: An analysis of composition and trends for the optimization of data collection. *Marine Policy,* 119, 104064.

Berto, Rita, and Giuseppe Barbiero (2017). How the psychological benefits associated with exposure to nature can affect pro-environmental behavior. *Annals of Cognitive Science,* 1 (1), 16–20.

Bonney, Rick, Heidi Ballard, Rebecca Jordan, Ellen McCallie, Tina Phillips, Jennifer Shirk, and Candie C. Wilderman (2009). *Public participation in scientific research: Defining the field and assessing its potential for informal science education.* A CAISE Inquiry Group Report. Washington, D.C. 20.03.2022 https://files.eric.ed.gov/fulltext/ED519688.pdf.

Boucher, Julien, and Damien Friot (2017). *Primary microplastics in the oceans: A global evaluation of sources.* IUCN International Union for Conservation of Nature. Gland.

Breakwell, Glynis M., Sean Hammond, Chris Fife-Schaw, and Jonathan A. Smith (eds.). (2006). *Research methods in psychology.* London: SAGE Publications, Ltd.

Catarino, Ana I., Johanna Kramm, Carolin Völker, Theodore B. Henry, and Gert Everaert (2021). Risk posed by microplastics: Scientific evidence and public perception. *Current Opinion in Green and Sustainable Chemistry,* 29, 100467.

Chen, Hongzhe, Sumin Wang, Huige Guo, Hui Lin, and Yuanbiao Zhang (2020). A nationwide assessment of litter on China's beaches using citizen science data. *Environmental Pollution,* 258, 113756.

Clayton, Susan, Christie M. Manning, Meighen Speiser, and Alison Nicole Hill (2017). *Mental health and our changing climate: Impacts, implications, and guidance.* American Psychological Association, and ecoAmerica.

Colmenero, Ana I., Claudio Barría, Elisabetta Broglio, and Salvador García-Barcelona (2017). Plastic debris straps on threatened blue shark *Prionace glauca*. *Marine Pollution Bulletin*, 115 (1–2), 436–38.

Conrad, Cathy C., and Krista G. Hilchey (2011). A review of citizen science and community-based environmental monitoring: Issues and opportunities. *Environmental Monitoring and Assessment*, 176 (1–4), 273–91.

Cooper, Caren B., Chris L. Hawn, Lincoln R. Larson, Julia K. Parrish, Gillian Bowser, Darlene Cavalier, Robert R. Dunn, Mordechai Haklay, Kaberi Kar Gupta, Na'Taki Osborne Jelks, Valerie A. Johnson, Madhusudan Katti, Zakiya Leggett, Omega R. Wilson, and Sacoby Wilson (2021). Inclusion in citizen science: The conundrum of rebranding. *Science*, 372 (6549), 1386–88.

Cowger, Win, Andrew B. Gray, and Richard C Schultz (2019). Anthropogenic litter cleanups in Iowa riparian areas reveal the importance of near-stream and watershed scale land use. *Environmental Pollution*, 250, 981–89.

Cronje, Ruth, Spencer Rohlinger, Alycia Crall, and Greg Newman (2011). Does participation in citizen science improve scientific literacy? A study to compare assessment methods. *Applied Environmental Education & Communication*, 10 (3), 135–45.

Davis, Wallace, and Anne G. Murphy (2015). Plastic in surface waters of the inside passage and beaches of the Salish Sea in Washington State. *Marine Pollution Bulletin*, 97 (1–2), 169–77.

Dempsey, Seraphim, Mel T. Devine, Tom Gillespie, Seán Lyons, and Anne Nolan (2018). Coastal blue space and depression in older adults. *Health & Place*, 54, 110–17.

Earn, Arielle, Kennedy Bucci, and Chelsea M Rochman (2021). A systematic review of the literature on plastic pollution in the Laurentian Great Lakes and its effects on freshwater biota. *Journal of Great Lakes Research*, 47 (1), 120–33.

ECSA (2015). Ten principles of citizen science. Berlin. 20.09.2021 https://doi.org/http://doi.org/10.17605/OSF.IO/XPR2N.

European Commission. 2020. Best practices in citizen science for environmental monitoring. 20.09.2021 https://ec.europa.eu/jrc/communities/en/community/examining-use-and-practices-citizen-science-eu-policies/page/best-practices-citizen.

Falk-Andersson, Jannike, Boris Woody Berkhout, and Tenaw Gedefaw Abate (2019). Citizen science for better management: Lessons learned from three Norwegian beach litter data sets. *Marine Pollution Bulletin*, 138, 364–75.

Forrest, Shaun A., Larissa Holman, Meaghan Murphy, and Jesse C. Vermaire (2019). Citizen science sampling programs as a technique for monitoring microplastic pollution: Results, lessons learned and recommendations for working with volunteers for monitoring plastic pollution in freshwater ecosystems. *Environmental Monitoring and Assessment*, 191 (3), 172.

Garcia-Soto, Carlos, Jan J.C. Seys, Oliver Zielinski, J. A. Busch, S. I. Luna, Jose Carlos Baez, C. Domegan, K. Dubsky, I. Kotynska-Zielinska, P. Loubat,

Francesca Malfatti, G. Mannaerts, Patricia McHugh, P. Monestiez, Gro I. van der Meeren, and G. Gorsky (2021). Marine citizen science: Current state in Europe and new technological developments. *Frontiers in Marine Science*, 8, 1–13.

GESAMP. 2019. Guidelines for the monitoring and assessment of plastic litter in the ocean. GESAMP Reports & Studies 99: 130. 20.09.2021 http://gesamp.org.

Gewert, Berit, Martin Ogonowski, Andreas Barth, and Matthew MacLeod (2017). Abundance and composition of near surface microplastics and plastic debris in the Stockholm Archipelago, Baltic Sea. *Marine Pollution Bulletin*, 120 (1–2), 292–302.

Grønhøj, Alice, and John Thøgersen (2012). Action speaks louder than words: The effect of personal attitudes and family norms on adolescents' pro-environmental behaviour. *Journal of Economic Psychology*, 33 (1), 292–302.

Grünzner, Maja, and Sabine Pahl (2022). Behavior change as part of the solution for plastic pollution. In Johanna Kramm and Carolin Völker (eds.). *Living in the plastic age*. Frankfurt, New York: Campus.

H2020 SOPHIE Consortium. 2020. A Strategic Research Agenda for Oceans and Human Health in Europe. Ostend, Belgium. 20.09.2021 https://doi.org/10.5281/zenodo.3696561.

Haklay, Mordechai Muki, Daniel Dörler, Florian Heigl, Marina Manzoni, Susanne Hecker, and Katrin Vohland (2021). What is citizen science? The challenges of definition. In Katrin Vohland, Anne Land-Zandstra, Luigi Ceccaroni, Rob Lemmens, Josep Perelló, Marisa Ponti, Roeland Samson, and Katherin Wagenknecht (eds.). *The science of citizen science*, 13–33. Cham: Springer International Publishing.

Haklay, Muki (2013). Citizen science and volunteered geographic information: Overview and typology of participation. In Daniel Sui, Sarah Elwood, and Michael Goodchild (eds.). *Crowdsourcing geographic knowledge: Volunteered geographic information (VGI) in theory and practice*, 105–122. Dordrecht: Springer Netherlands.

Harris, Lucas, Max Liboiron, Louis Charron, and Charles Mather (2021). Using citizen science to evaluate extended producer responsibility policy to reduce marine plastic debris shows no reduction in pollution levels. *Marine Policy*, 123, 104319.

Hartley, Bonny L., Sabine Pahl, Joana Veiga, Thomais Vlachogianni, Lia Vasconcelos, Thomas Maes, Tom Doyle, Ryan d'Arcy Metcalfe, Ayaka Amaha Öztürk, Mara Di Berardo, and Richard C. Thompson (2018). Exploring public views on marine litter in Europe: Perceived causes, consequences and pathways to change. *Marine Pollution Bulletin*, 133, 945–55.

Hartley, Bonny L., Richard C. Thompson, and Sabine Pahl (2015). Marine litter education boosts children's understanding and self-reported actions. *Marine Pollution Bulletin*, 90 (1–2), 209–17.

Heigl, Florian, Barbara Kieslinger, Katharina T. Paul, Julia Uhlik, and Daniel Dörler (2019). Opinion: Toward an international definition of citizen science. *Proceedings of the National Academy of Sciences*, 116 (17), 8089–8092.

Hidalgo-Ruz, Valeria, and Martin Thiel (2013). Distribution and abundance of small plastic debris on beaches in the SE Pacific (Chile): A study supported by a citizen science project. *Marine Environmental Research*, 87–88, 12–18.

Hidalgo-Ruz, Valeria, and Martin Thiel (2015). The contribution of citizen scientists to the monitoring of marine litter. In Melanie Bergmann, Lars Gutow, and Michael Klages (eds.). *Marine anthropogenic litter*, 429–47. Cham: Springer International Publishing.

Hines, Jody M., Harold R. Hungerford, and Audrey N. Tomera (1987). Analysis and synthesis of research on responsible environmental behavior: A meta-analysis. *The Journal of Environmental Education*, 18 (2), 1–8.

Honorato-Zimmer, Daniela, Katrin Kruse, Katrin Knickmeier, Anna Weinmann, Ivan A Hinojosa, and Martin Thiel (2019). Inter-hemispherical shoreline surveys of anthropogenic marine debris – A binational citizen science project with schoolchildren. *Marine Pollution Bulletin*, 138, 464–73.

Jambeck, Jenna R., Roland Geyer, Chris Wilcox, Theodore R. Siegler, Miriam Perryman, Anthony Andrady, Ramani Narayan, and Kara Lavender Law (2015). Plastic waste inputs from land into the ocean. *Science*, 347 (6223), 768–71.

Kaida, Naoko, and Kosuke Kaida (2016). Facilitating pro-environmental behavior: The role of pessimism and anthropocentric environmental values. *Social Indicators Research*, 126 (3), 1243–1260.

Kelly, Rachel, Karen Evans, Karen Alexander, Silvana Bettiol, Stuart Corney, Coco Cullen-Knox, Christopher Cvitanovic, Kristy de Salas, Gholam Reza Emad, Liam Fullbrook, Carolina Garcia, Sierra Ison, Scott Ling, Catriona Macleod, Amelie Meyer, Linda Murray, Michael Murunga, Kirsty L. Nash, Kimberley Norris, Michael Oellermann, Jennifer Scott, Jonathan S. Stark, Graham Wood, and Gretta T. Pecl (2022). Connecting to the oceans: Supporting ocean literacy and public engagement. *Reviews in Fish Biology and Fisheries*, 32 (1), 123–143.

Kiessling, Tim, Katrin Knickmeier, Katrin Kruse, Dennis Brennecke, Alice Nauendorf, and Martin Thiel (2019). Plastic pirates sample litter at rivers in Germany – Riverside litter and litter sources estimated by schoolchildren. *Environmental Pollution*, 245, 545–57.

Kiessling, Tim, Sonia Salas, Konar Mutafoglu, and Martin Thiel (2017). Who cares about dirty beaches? Evaluating environmental awareness and action on coastal litter in Chile. *Ocean & Coastal Management*, 137, 82–95.

Lau, Winnie W. Y., Yonathan Shiran, Richard M. Bailey, Ed Cook, Martin R. Stuchtey, Julia Koskella, Costas A. Velis, Linda Godfrey, Julien Boucher, Margaret B. Murphy, Richard C. Thompson, Emilia Jankowska, Arturo Castillo Castillo, Toby D. Pilditch, Ben Dixon, Laura Koerselman, Edward Kosior, Enzo Favoino, Jutta Gutberlet, Sarah Baulch, Meera E. Atreya, David Fischer, Kevin K. He, Milan M. Petit, U. Rashid Sumaila, Emily Neil, Mark V. Bernhofen, Keith Lawrence, and James E. Palardy (2020). Evaluating scenarios toward zero plastic pollution. *Science*, 369 (6510), 1455–1461.

Lee, Jongmyoung, Sunwook Hong, and Jongsu Lee (2019). Rapid assessment of marine debris in coastal areas using a visual scoring indicator. *Marine Pollution Bulletin*, 149, 110552.

Lippiatt, Sherry, Sarah Opfer, and Courtney Arthur (2013). Marine Debris Monitoring and Assessment: Recommendations for Monitoring Debris Trends in the Marine Environment. NOAA Technical Memorandum, no. NOS-OR&R-46: 88. 20.09.2021 http://marinedebris.noaa.gov/sites/default/files/Lippiatt_et_al_2013.pdf.

Locritani, Marina, Silvia Merlino, and Marinella Abbate (2019). Assessing the citizen science approach as tool to increase awareness on the marine litter problem. *Marine Pollution Bulletin*, 140, 320–29.

Loizidou, Xenia I., Michael I. Loizides, and Demetra L. Orthodoxou (2018). Persistent marine litter: Small plastics and cigarette butts remain on beaches after organized beach cleanups. *Environmental Monitoring and Assessment*, 190 (7), 414.

Lots, Froukje A. E., Paul Behrens, Martina G. Vijver, Alice A. Horton, and Thijs Bosker (2017). A large-scale investigation of microplastic contamination: abundance and characteristics of microplastics in European beach sediment. *Marine Pollution Bulletin*, 123 (1–2), 219–26.

Lucrezi, Serena, and Oghenetejiri Digun-Aweto (2020). 'Who wants to join?' Visitors' willingness to participate in beach litter clean-ups in Nigeria. *Marine Pollution Bulletin*, 155, 111167.

Maguire, Grainne S., Kelly K. Miller, Michael A. Weston, and Kirsten Young (2011). Being beside the seaside: Beach use and preferences among coastal residents of South-Eastern Australia. *Ocean & Coastal Management*, 54 (10), 781–88.

Mayoma, Bahati S., Innocent S. Mjumira, Aubrery Efudala, Kristian Syberg, and Farhan R. Khan (2019). Collection of anthropogenic litter from the shores of Lake Malawi: Characterization of plastic debris and the implications of public involvement in the African Great Lakes. *Toxics*, 7 (4), 64.

Mioni, E., S. Merlino, M. Locritani, S. Strada, A. Giovacchini, M. Stroobant, and R. Traverso (2015). 'Blue Paths' and SEACleaner: Ensuring long-term commitment of citizens in environmental monitoring and scientific research. OCEANS 2015 – Genova. 20.09.2021 https://doi.org/10.1109/OCEANS-Genova.2015.7271666.

Moore, Charles James (2008). Synthetic polymers in the marine environment: A rapidly increasing, long-term threat. *Environmental Research*, 108 (2), 131–39.

Nelms, Sarah E., Charlotte Coombes, Laura C Foster, Tamara S. Galloway, Brendan J Godley, Penelope K. Lindeque, and Matthew John Witt (2017). Marine anthropogenic litter on British beaches: A 10-year nationwide assessment using citizen science data. *Science of the Total Environment*, 579, 1399–1409.

Nelms, Sarah E., Lauren Eyles, Brendan J. Godley, Peter B. Richardson, Hazel Selley, Jean-Luc Solandt, and Matthew J. Witt (2020). Investigating the distribution and regional occurrence of anthropogenic litter in English marine protected areas using 25 years of citizen-science beach clean data. *Environmental Pollution*, 263, 114365.

OECD (2006). *Assessing scientific, reading and mathematical literacy*. Pisa: OECD. 20.09.2021 https://doi.org/10.1787/9789264026407-en.

Page, Matthew J., Joanne E. McKenzie, Patrick M. Bossuyt, Isabelle Boutron, Tammy C. Hoffmann, Cynthia D. Mulrow, Larissa Shamseer, Jennifer M. Tetzlaff, Elie A. Akl, Sue E. Brennan, Roger Chou, Julie Glanville, Jeremy M. Grimshaw, Asbjørn Hróbjartsson, Manoj M. Lalu, Tianjing Li, Elizabeth W. Loder, Evan Mayo-Wilson, Steve McDonald, Luke A. McGuinness, Lesley A. Stewart, James Thomas, Andrea C. Tricco, Vivian A. Welch, Penny Whiting, and David Moher (2021). The PRISMA 2020 statement: An updated guideline for reporting systematic reviews. *BMJ*, 372, n71.

Pandya, Rajul, and Kenne Ann Dibner (eds.). (2018). *Learning through citizen science: Enhancing opportunities by design*. Washington, D.C.: National Academies Press. https://doi.org/10.17226/25183.

Pandya, Rajul E. (2012). A framework for engaging diverse communities in citizen science in the US. *Frontiers in Ecology and the Environment*, 10 (6), 314–17.

Partnership for Observation of the Global Ocean (2021). COLLECT – Citizen Observation of Local Litter in Coastal ECosysTems. 2021. https://pogo-ocean.org/innovation-in-ocean-observing/activities/collect-citizen-observation-of-local-litter-in-coastal-ecosystems/.

Pateman, Rachel, Alison Dyke, and Sarah West (2021). The diversity of participants in environmental citizen science. *Citizen Science: Theory and Practice*, 6 (1), 1–16.

Peng, Chenchen, and Kazuo Yamashita (2016). Effects of the coastal environment on well-being. *Journal of Coastal Zone Management*, 19 (2), 1–7.

Peter, Maria, Tim Diekötter, Tim Höffler, and Kerstin Kremer (2021). Biodiversity citizen science: Outcomes for the participating citizens. *People and Nature*, 3 (2), 294–311.

Pihkala, Panu (2020). Eco-anxiety and environmental education. *Sustainability*, 12 (23), 10149.

R Core Team (2020). *R: A language and environment for statistical computing*. Vienna, Austria: R Foundation for Statistical Computing. https://www.r-project.org.

Rambonnet, Liselotte, Suzanne C. Vink, Anne M. Land-Zandstra, and Thijs Bosker (2019). Making citizen science count: Best practices and challenges of citizen science projects on plastics in aquatic environments. *Marine Pollution Bulletin*, 145, 271–77.

Rayon-Viña, Fernando, Laura Miralles, Sara Fernandez-Rodríguez, Eduardo Dopico, and Eva Garcia-Vazquez (2019). Marine litter and public involvement in beach cleaning: Disentangling perception and awareness among adults and children, Bay of Biscay, Spain. *Marine Pollution Bulletin*, 141, 112–18.

Rayon-Viña, Fernando, Laura Miralles, Marta Gómez-Agenjo, Eduardo Dopico, and Eva Garcia-Vazquez (2018). Marine litter in South Bay of Biscay: Local differences in beach littering are associated with citizen perception and awareness. *Marine Pollution Bulletin*, 131, 727–35.

Roman, Lauren, Britta Denise Hardesty, George H. Leonard, Hannah Pragnell-Raasch, Nicholas Mallos, Ian Campbell, and Chris Wilcox (2020). A global assessment of the relationship between anthropogenic debris on land and the seafloor. *Environmental Pollution*, 264, 114663.

Sea Change (2018). The sea and our physical and mental wellbeing. 2018. 20.09.2021 https://www.worldoceannetwork.org/wp-content/uploads/2016/09/5.-Blue-Gym.pdf.

Sheppard, James (2004). Reducing pessimism's sway in the environmental ethics classroom. *Worldviews: Global Religions, Culture, and Ecology*, 8 (2–3), 213–26.

Smith, Stephen D. A., Kelsey Banister, Nicola Fraser, and Robert J. Edgar (2018). Tracing the source of marine debris on the beaches of Northern New South Wales, Australia: The Bottles on Beaches Program. *Marine Pollution Bulletin*, 126, 304–307.

Strang, Craig, Annette DeCharon, and Sarah Schoedinger (2007). Can you be science literate without being ocean literate? *The Journal of Marine Education*, 23 (1), 7–9.

Syberg, Kristian, Annemette Palmqvist, Farhan R. Khan, Jakob Strand, Jes Vollertsen, Lauge Peter Westergaard Clausen, Louise Feld, Nanna B. Hartmann, Nikoline Oturai, Søren Møller, Torkel Gissel Nielsen, Yvonne Shashoua, and Steffen Foss Hansen (2020). A nationwide assessment of plastic pollution in the Danish realm using citizen science. *Scientific Reports*, 10 (1), 1–11.

Tauginienė, Loreta, Eglė Butkevičienė, Katrin Vohland, Barbara Heinisch, Maria Daskolia, Monika Suškevičs, Manuel Portela, Bálint Balázs, and Baiba Prūse (2020). Citizen science in the social sciences and humanities: The power of interdisciplinarity. *Palgrave Communications*, 6 (1), 89.

Taylor, George R. (ed.) (2005). *Integrating quantitative and qualitative methods in research*. Lanham, Boulder, New York, Toronto, Oxford: University Press of America.

Tekman, Mine B., Lars Gutow, Ana Macario, Antonie Haas, Andreas Walter, and Melanie Bergmann (2021). *LITTERBASE*. Bremerhaven: AWI. Alfred-Wegener-Institut Helmholtz-Zentrum für Polar- und Meeresforschung. https://litterbase.awi.de/interaction_detail

The World Bank (2021). World Bank Country and Lending Groups. 2021. https://datahelpdesk.worldbank.org/knowledgebase/articles/906519-world-bank-country-and-lending-groups.

Toomey, Anne H., and Margret C. Domroese (2013). Can citizen science lead to positive conservation attitudes and behaviors? *Research in Human Ecology*, 20 (1), 50–62.

Triguero-Mas, Margarita, David Donaire-Gonzalez, Edmund Seto, Antònia Valentín, David Martínez, Graham Smith, Gemma Hurst, Glòria Carrasco-Turigas, Daniel Masterson, Magdalena van den Berg, Albert Ambròs, Tania Martínez-Iñiguez, Audrius Dedele, Naomi Ellis, Tomas Grazulevicius, Martin Voorsmit, Marta Cirach, Judith Cirac-Claveras, Wim Swart, Eddy Clasquin,

Annemarie Ruijsbroek, Jolanda Maas, Michael Jerret, Regina Gražulevičienė, Hanneke Kruize, Christopher J. Gidlow, and Mark J. Nieuwenhuijsen (2017). Natural outdoor environments and mental health: Stress as a possible mechanism. *Environmental Research*, 159, 629–38.

Trumbull, Deborah J., Rick Bonney, Derek Bascom, and Anna Cabral (2000). Thinking scientifically during participation in a citizen-science project. *Science Education*, 84 (2), 265–75.

Tunnell, Jace W., Kelly H. Dunning, Lindsay P. Scheef, and Kathleen M. Swanson (2020). Measuring Plastic Pellet (Nurdle) Abundance on shorelines throughout the Gulf of Mexico using citizen scientists: Establishing a platform for policy-relevant research. *Marine Pollution Bulletin*, 151, 110794.

Turner, Andrew, Tracey Williams, and Tom Pitchford (2021). Transport, weathering and pollution of plastic from container losses at sea: Observations from a spillage of inkjet cartridges in the North Atlantic Ocean. *Environmental Pollution*, 284, 117131.

Turrell, William R. (2019). Spatial distribution of foreshore litter on the Northwest European continental shelf. *Marine Pollution Bulletin*, 142, 583–94.

Uhrin, Amy V., Sherry Lippiatt, Carlie E. Herring, Kyle Dettloff, Kate Bimrose, and Chris Butler-Minor (2020). Temporal trends and potential drivers of stranded marine debris on beaches within two US national marine sanctuaries using citizen science data. *Frontiers in Environmental Science*, 8. 604927.

UNEP (2014). *Valuing plastics: The business case for measuring, managing and disclosing plastic use in the consumer goods industry*. Nairobi: United Nations Environment Programme (UNEP). 20.09.2021 https://wedocs.unep.org/bitstream/handle/20.500.11822/25302/Valuing_Plastic_ES.pdf?sequence=1&isAllowed=y%0A www.gpa.unep.org%0Awww.unep.org/pdf/ValuingPlastic/.

van der Velde, Tonya, David A. Milton, T. J. Lawson, Chris Wilcox, Matt Lansdell, Geraldine Davis, Genevieve Perkins, and Britta Denise Hardesty (2017). Comparison of marine debris data collected by researchers and citizen scientists: Is citizen science data worth the effort? *Biological Conservation*, 208, 127–38.

van Hee, Francine M., Arya Seldenrath, and Jan Seys (2020). *Policy informing brief: Marine citizen science in the North Sea area and what policy makers can learn from it*. Oostend: Flanders Marine Institute (VLIZ).

Walther, Bruno A., Alexander Kunz, and Chieh-Shen Hu (2018). Type and quantity of coastal debris pollution in Taiwan: A 12-year nationwide assessment using citizen science data. *Marine Pollution Bulletin*, 135, 862–72.

West, Sarah, Alison Dyke, and Rachel Pateman (2021). Variations in the motivations of environmental citizen scientists. *Citizen Science: Theory and Practice*, 6 (1), 1–18.

Western Indian Ocean Marine Science Association (2021). WIOMSA marine litter monitoring project. 20.09.2021 https://sst.org.za/projects/african-marine-waste-network/wiomsa-marine-litter-monitoring-project/.

Westfall, Raine, and Grace Simantel (2019). Public views on ocean literacy and plastic pollution in Washington Harbor towns. 20.09.201. https://wp.wwu.edu/oceanliteracy/findings/.

Wheeler, Benedict W., Mathew White, Will Stahl-Timmins, and Michael H. Depledge (2012). Does living by the coast improve health and wellbeing? *Health & Place*, 18 (5), 1198–1201.

White, Mathew P., Ian Alcock, Benedict W. Wheeler, and Michael H. Depledge (2013). Coastal proximity, health and well-being: results from a longitudinal panel survey. *Health & Place*, 23, 97–103.

Wickham, Hadley (2016). *Ggplot2: Elegant graphics for data analysis*. New York: Springer-Verlag.

Wyles, Kayleigh J., Sabine Pahl, Matthew Holland, and Richard C. Thompson (2017). Can beach cleans do more than clean-up litter? Comparing beach cleans to other coastal activities. *Environment and Behavior*, 49 (5), 509–35.

Wyles, Kayleigh J., Sabine Pahl, and Richard C. Thompson (2014). Perceived risks and benefits of recreational visits to the marine environment: Integrating impacts on the environment and impacts on the visitor. *Ocean & Coastal Management*, 88, 53–63.

Wyles, Kayleigh J., Sabine Pahl, Katrina Thomas, and Richard C. Thompson (2016). Factors that can undermine the psychological benefits of coastal environments. *Environment and Behavior*, 48 (9), 1095–1126.

Yeo, Bee Geok, Hideshige Takada, Heidi Taylor, Maki Ito, Junki Hosoda, Mayumi Allinson, Sharnie Connell, Laura Greaves, and John McGrath (2015). POPs monitoring in Australia and New Zealand using plastic resin pellets, and International Pellet Watch as a tool for education and raising public awareness on plastic debris and POPs. *Marine Pollution Bulletin*, 101 (1), 137–45.

Zettler, Erik R., Hideshige Takada, Bonnie Monteleone, Nicholas J. Mallos, Marcus Eriksen, and Linda A. Amaral-Zettler (2017). Incorporating citizen science to study plastics in the environment. *Analytical Methods*, 9 (9), 1392–1403.

APPENDIX A

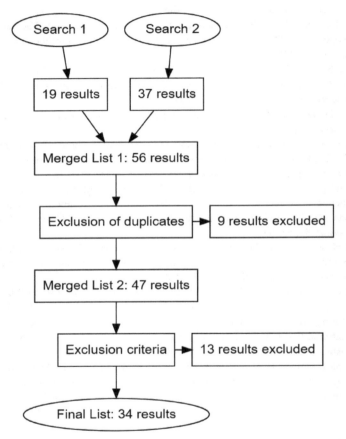

Preferred Reporting Items for Systematic Reviews and Meta-Analyses (PRISMA) flow diagram [for search terms please refer to Tab. 1, for exclusion criteria please see main text]. Results were retrieved via Scopus (Elsevier) on 19/05/2021 and merged lists were managed using Mendeley (Elsevier). The final list of results consisted of 34 scientific peer-reviewed articles is available in Tab. 2.

Behavior Change as Part of the Solution for Plastic Pollution

Maja Grünzner, Sabine Pahl

Introduction

Problem awareness of the public and risk perception are important elements to understand people's mental representation of a hazard, in understanding the "status quo", of current beliefs and opinions (for more on risk perception see Felipe-Rodriguez et al. in this volume). Nevertheless, further actions need to follow if we want to change the "status quo" and encourage environmentally friendly behaviors. Therefore, the following chapter aims to provide an overview of how social and environmental psychology approaches focusing on behavior change can be applied to plastic pollution and in this way contribute to solutions. To clarify, environmental psychology is defined "as the discipline that studies the interplay between individuals and the built and natural environment" (Steg et al. 2018, 2). This chapter will introduce different theoretical models and approaches of behavior change, followed by techniques used in the context of consumer-focused interventions. Moreover, we summarize potential barriers towards environmentally friendly behaviors known as "Dragons of Inaction". Lastly, we emphasize the importance of intervention development, evaluation, and communication techniques firmly based on scientific evidence from social and environmental psychology. This chapter does not present a complete literature review on what has been done (for reviews see Heidbreder et al. 2019 and Nuojua et al. 2021). Instead, it aims to explain how interventions reducing plastic consumption can be developed using theories from psychology.

Exploring the Role of Environmental Psychology in the Plastic Discourse

Human activities are the sole cause of plastic pollution in our natural environment (Pahl and Wyles 2017); hence, behavior change can be one solution to tackle plastic pollution. Behavioral approaches are on the rise but still underrepresented in the plastics discourse (Nuojua et al. 2021; SAPEA 2019). Investigating behavior and its antecedents from an environmental psychology perspective is also gaining importance in other environmental contexts such as nature conservation and climate change. However, the positive impact of environmental psychology knowledge about climate change action (Steg 2018; van Valkengoed and Steg 2019) and conservation efforts (Nielsen et al. 2021; Schultz 2011) has not reached its full potential (Nielsen et al. 2021; Whitmarsh et al. 2021).

One way forward is to focus on environmentally impactful behaviors understood as individual behaviors which have a significant influence on changing the structure or dynamics in our environment (Stern 2000) and to do so systematically (Nielsen et al. 2021), using qualitative and quantitative research.

This is a useful starting point for this chapter and more broadly for scientists working on tackling plastic pollution. Plastic production, consumption and disposal are all influenced by human behavior from different stakeholders at different levels (Pahl et al. 2020). Therefore, we hope that the relevance of environmental psychology continues to become more apparent in the wider scientific discourse and across disciplines. We suggest a focus on the potentially riskiest plastics and behaviors as well as underlying human factors, which are relevant for that specific plastic being produced, used, and disposed (see also SAPEA 2019). Interdisciplinary collaborations are therefore necessary to understand the plastic system fully. The technical life-cycle assessment of a problematic plastic product, for example, developed by environmental engineers will be useful for environmental psychologists wanting to explore predictors of behavior to develop communication and behavior change interventions. Whereas, the research informed interventions will be beneficial for governments and municipalities on a local, national or international level to have a positive real-life impact for its communities such as a decrease in plastic pollution. The EU-funded H2020 LimnoPlast project, for instance, investigates microplastics pollution and solutions in Europe's freshwater ecosystems from an environmental, technical, and societal perspective (https://www.limnoplast-itn.eu/). The psychologists in the project investigate the perceptions of experts and laypeople to work towards suitable risk communication and behavioral change measures; all under consideration of current research insights about psychological and contextual constraints. This is crucial for the progress towards the reduction of plastic pollution, thus we interweave these in the remainder of the chapter, starting with a discussion of the awareness-behavior gap.

The Awareness-Behavior Gap

Awareness-raising among citizens—an often-proposed action against plastic pollution—is an important part to tackle plastic pollution. However, it is not the complete answer to this problem. Europeans, for instance, are aware of the plastic problem and express concern regarding a range of impacts (Davison et al. 2021; European Commission 2020); however, research has shown that problem awareness is only one of many variables influencing sustainable behavior (Heidbreder et al. 2019; van Valkengoed and Steg 2019).

A Small Thought Experiment:

> Imagine you are being invited to a barbecue with a few friends at a public barbecue place. The sun is shining and your work for today is done. There is only one more thing to do—buying a bottle of wine or non-alcoholic beverage from the supermarket and bringing cutlery for everyone. Now you have the choice to bring the stainless steel cutlery you have at home or to buy the single-use plastic cutlery at the supermarket. Both options have their advantages and disadvantages.
> What do you bring and why?

Many different factors appear to influence people's choice of reusable cutlery or single-use cutlery. We are going to explore one of the factors influencing this decision below because being aware of the negative impact of, in this case, single-use plastic will probably not be the main factor determining the choice.

Heidbreder et al. (2019) identified a gap between the awareness of plastic pollution and related behaviors, that is, even if people know about the negative impacts of plastic pollution, they do not always act accordingly (see also Stieß et al. in this volume). Instead, their behaviors are mainly predicted by habits, social, and situational factors. A potential explanation of why awareness alone will not change environmental behavior can be found in the norm-activation model (Schwartz 1977; Schwartz and Howard 1981). The norm-activation model, originally developed for explaining altruistic behavior assumes that a feeling of moral obligations (personal norm) influences people's actions in moral situations (Schwartz 1977; Schwartz and Howard 1981). To do so, the personal norm—understood as the representation of personal values in the present moral situation—needs to be activated to trigger the action (Schwartz 1977; Schwartz and Howard 1981).

Personal norms will be activated in people when 1) they become aware of the need that someone or something needs help (awareness of need), 2) they become aware that certain actions have consequences—that these can increase or decrease the problem (awareness of consequences), 3) they ascribe personal responsibility to the problem (ascription of responsibility), 4) they feel capable of helping or doing something against the problem (self-efficacy; Schwartz and Howard 1981).

Even though people are typically aware of the consequences when they are buying single-use cutlery, it is not given that they will choose the latter. It could be that the person does not feel responsible for global plastic pollution, rather ascribing the responsibility mainly to the producers, retailers or broadly to industry instead of themselves. It also could be that the person does not feel capable of acting environmentally friendly in the situation, which will hinder the activation of the moral obligation and make the purchase of single-use plastic cutlery more likely.

To summarize, personal norms—as the moral influence on acting in an environmentally friendly manner—need to be activated in people by different factors, one of them being aware of needs and consequences.[1] Awareness does play a role but is only one factor of many and often has no direct influence on behavior. That means if behavior change is the aim, raising awareness of need and consequences should be accompanied by information about individual responsibility and information about how to act (morally) "right" or socially appropriate in the given situation.

Behavior Change: From Theory to Practice

Personal norms are not the only behavioral motivator. In case a practitioner (e.g., a municipality managing a public barbecue area or local waste

---

1 Due to the similarity of *awareness of need* and *awareness of consequences* multiple researchers have adapted the model and been only using one of the constructs (Klöckner 2013a).

collection) would like to create an intervention to decrease people bringing single-use plastics to the area, it helps to understand why people use single-use plastics in the first place. Therefore, in this section, we would like to dive deeper into the mechanisms of behavior and behavior change. We introduce two currently used and fairly extensive models as examples—the *comprehensive action determination model* and the *stage model of self-regulated behavioral change*. In this chapter, they are guiding the exploration of behavioral determinants and we are using them as adaptable frameworks.

Prediction Model: The Comprehensive Action Determination Model

Various theories and models have been used to predict environmental behavior outcomes and to investigate possibilities for transforming environmental behavior and its motivators. However, many theories reduced the prediction of behavior to a few of these motivators. Some focused on normative, some on non-normative motivators. Environmental psychologists (Klöckner 2013a; 2015; Klöckner and Blöbaum 2010), therefore, concluded that a model with normative and non-normative motivators was needed and joined three commonly used models: The theory of planned behavior (Ajzen 1991), the norm activation model (Schwartz 1977; Schwartz and Howard 1981) and the value-belief-norm theory (Stern 2000). Based on these theories the comprehensive action determination model was developed (Klöckner 2013b; see Fig. 1). The theories complement each other, as the often-criticized missing morality construct in the theory of planned behavior can be implemented when considering variables from the norm activation model and value-belief-norm theory (Klöckner 2013a). Additionally, none of the original theories explains repetitive behavior well and therefore, habit strength was included in the comprehensive action determination model. It is empirically supported and has been tested with different environmental behaviors (Klöckner 2013a). The comprehensive action determination model focuses on intrapersonal constructs, but also includes social and situational influences. As mentioned above, the intrapersonal constructs are divided into normative and non-normative parts (also defined as moral and nonmoral) which influence each other and can directly or indirectly motivate behavior. Values, ecological worldview, awareness of consequences and ascription to responsibility indirectly motivate behavior, therefore, their influence can be interrupted or weakened by various factors such as competing attitudes, low perceived behavioral control etc. (see Fig. 1). Nevertheless, they play an important role as they, together with social norms, create moral obligations (Klöckner 2013a). Moreover, personal norms, social norms, attitudes, and perceived behavioral control influence behavioral intention, whereas intention, perceived behavioral control, and habits influence behavior. They are the direct behavioral motivators (Klöckner 2013a).

Fig. 1: The comprehensive action determination model. (Source: Adapted from "How powerful are moral motivations in environmental protection?: An integrated model framework." by C.A. Klöckner 2013. In Handbook of moral motivation, p. 462.

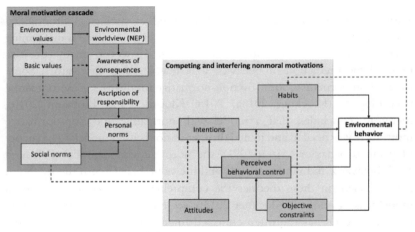

Copyright 2013 by Sense Publishers)

The values specifically studied in the context of environmental behavior and therefore, relevant for the comprehensive action determination model, are biospheric, altruistic, and egoistic values. Biospheric and altruistic values are defined in terms of concern towards nature (biospheric) and human wellbeing (altruistic), respectively, whereas egoistic values are defined in terms of concern about or interest in oneself (Stern 2000; Stern and Dietz 1994). Values are linked with ecological worldview (a measure created by Dunlap et al. 2000). The ecological worldview contains the beliefs that human behaviors threaten nature, that natural resources are finite and that humans should not rule over nature (Dunlap et al. 2000). People with strong biospheric and altruistic values are likely to have a more ecological worldview while people with strong egoistic values are likely to have a less ecological worldview (Klöckner 2013a). Based on the value-belief-norm theory, the ecological worldview takes its place as a link between values and personal norms instead of its classical interpretation as an environmental attitude (Stern 2000).

Norms are the shared beliefs of a person on how they should act. They can be categorized by how externalized or internalized they are. Internalized norms such as personal norms are understood as "the self-expectations for specific action in particular situations that are constructed by the individual." (Schwartz 1977, 227). As described in the previous sections, personal norms need to be activated and are expressed as "feelings of moral obligation" (Schwartz 1977). On the other hand, social norms are external and broadly described as the perceived social pressure to perform a behavior relevant to other people (and reinforced by rewards or punishments) in a certain situation (Schwartz 1977). However, a study about recycling and organic food purchase with Portuguese and Brazilian participants showed that personal norms predicted the environmentally friendly behaviors better than social norms (Bertoldo and Castro 2016).

All the constructs described above are normative motivations to act environmentally friendly. The great strength in the comprehensive action determination models lays in its wide scope—combining normative motivations with non-normative motivations to predict environmentally friendly behavior. The latter include attitudes and perceived behavioral control. Based on Ajzen's theory of Planned Behavior (1991), attitudes are understood as the degree of the person's evaluation (favorable or unfavorable) towards a specific behavior and its outcomes, whereas, perceived behavioral control is understood as the perceived ease or difficulty of an action. It is manifested by past experiences and perceived anticipated barriers (Ajzen 1991). According to the comprehensive action determination model, all of these factors influence behavior in diverse ways and therefore, need to be targeted differently (Klöckner 2015).

Moreover, the comprehensive action determination model includes habit strength (more details found in Box 1) as a predictor for repeating behaviors, as habit strength moderates the relationship between the intention and the actual behavior. The stronger the habit, the weaker the influence of intention on behavior (Klöckner 2013a).

> Box 1: Habitual Behavior and Plastic Consumption
>
> Habits are a key predictor of plastic consumption and they are closely interlinked with context (Heidbreder et al. 2019). One example in regards to plastics is a study by Romero et al. (2018) with Brazilian immigrants in Canada. In this study, participants compared their old environmental attitudes and behaviors in Brazil to their present attitudes and behaviors using self-assessment. The participants reported that they already felt pro-environmental before moving and that the use of plastic bags was "just a habit" (Romero et al. 2018, 8) of them in their home country. After they migrated to Canada, the participants reduced plastic bag usage while no environmental attitude change was reported. Moreover, this fits with the assumption that individuals only reconsider their habits when the context changes drastically (Steg and Vlek 2009).
> Furthermore, habits are woven with situational factors such as convenience. A study with participants from South Africa showed "convenience" (51 percent) was chosen as the main reason for plastic bag usage. In comparison, forgetting to bring their reusable bag was selected by 13 percent of the participants (O'Brien and Thondhlana 2019). This is in line with past research finding similar results regarding respondents using plastic bags out of convenience (Braun and Traore 2015), next to using plastic bags because of easy access or their low price (Adane and Muleta 2011; O'Brien and Thondhlana 2019).
> Heidbreder et al. (2019) point out that social interventions (political and psychological) seem to also be potentially effective for habits. For example, a voice prompt from cashiers (asking if customers want a free plastic bag in Japanese supermarkets) led to a decrease of plastic bag usage of 5 percent, in comparison to cashiers handing out plastic bags to the customers without asking (Ohtomo and Ohnuma 2014). One study in Portugal reported that avoiding payments of a plastic carrier bag tax was one of the main reasons for not using plastic bags in the short and medium term (Martinho et al. 2017). Moreover, after the implementation of the charge, Portuguese participants from another study reported developing reuse habits and a reduction of single-use plastic usage (Luís et al. 2020). Nevertheless, long-term changes need to be investigated further (Heidbreder et al. 2019).

Generally, habits fulfil a function to achieve specific goals, they are automatic behavioral patterns and bound to a certain stable situation

(Verplanken and Aarts 1999). They develop through positive reinforcement (e.g., achieving the situational specific goal) of the repeated behavior.

Example: Reducing Personal Clothing Consumption

Synthetic microfibers are one of the major microplastics sources in European rivers (Siegfried et al. 2017), plus experts working on plastics perceive the impact of textile microfibers as one of the riskiest on the natural environment and human health in comparison to other sources, such as bigger plastic items breaking down (Grünzner et al. 2021). Moreover, fast fashion—which is mainly using synthetic fibers and harmful chemicals—has taken over the clothing market (Niinimäki et al. 2020); hence, people's reduction in their clothing purchases can contribute to decreasing microplastics pollution.

Joanes et al. (2020) applied the comprehensive action determination model to determine the most influential intrapersonal factors for reducing clothing consumption. This is highly valuable as pathways for behavioral interventions can be identified. The results from their two studies show that across five countries (Germany, Poland, Sweden, United States and United Kingdom; n = 5,185), personal norms and social norms were the strongest predictors for intention to reduce clothing consumption, followed by attitudes. Norms having the strongest effect led the authors to the assumption that clothing purchase behavior has moral components. Perceived behavioral control had a negative relationship with purchase behavior (measured daily, during a two-week period). That means that the more participants felt capable to reduce clothing purchases, the least they bought. Perceived behavioral control was rated high across all countries and the authors suspect that this could be due to the easy nature of the consumption behavior. Consumers are theoretically able to reduce clothing purchases right away. Nonetheless, intention to reduce clothing consumption was not related to perceived behavioral control. Moreover, the authors included a measure of impulsive purchase behavior and past behavior. Impulsive purchase behavior was related to past as well as actual purchase behavior, which made the authors conclude that the behavior in question can be non-intentional or potentially even automatic sometimes.

Several potential intervention approaches are proposed by Joanes and their colleagues (2020): 1) Increasing (and activating) personal norms through wide-spread information campaigns about the environmental impact of clothes, such as where and how much pollution it causes; providing specific groups with information about how resources can be saved when reducing one's clothing purchase; increasing awareness about the impact of individual clothing consumption. 2) Instructing consumers to reflect on past purchases to make them think about unnecessary purchases and potentially influence future behavior. 3) Exploring strategies based on self-regulation theory including goal setting, implementation intentions or if-then plans to tackle the intention-behavior gap to increase the success rate of the action or in this case non-action e.g., no purchase (Bamberg 2013; Nielsen 2017; Sheeran and Webb 2016).

Stage Model: The Stage Model of Self-Regulated Behavioral Change

Following up on our last example of clothing purchases, we want to introduce the intention-behavior gap and explain why applying the stage model of self-regulated behavioral change can help to increase the success of translating the behavioral intention into doing the behavior. The study from Joanes et al. (2020) found that the higher a participant scored on the importance of the goal to reduce clothing consumption, the fewer clothing items were purchased. Nonetheless, the effect was small which the authors suspect could be related to the frequently observed intention-behavior gap. A reason for this could be that prediction models such as the comprehensive action determination model frame behavior change as a static process at one point of time even though behavior change often occurs over a longer time period (Gollwitzer 1990).

In response to this issue, Bamberg (2013) suggested applying a stage approach—studying people's voluntary change—which has been shown to be successful in targeting behaviors related to health (see e.g., Schwarzer 2008). The stage model of self-regulated behavioral change (Bamberg 2013, Fig. 2) integrates the model of action phases (Gollwitzer 1990) and determinants from the norm-activation model (Schwartz 1977; Schwartz and Howard 1981), theory of planned behavior (Ajzen 1991) and health action process approach (Schwarzer 2008).

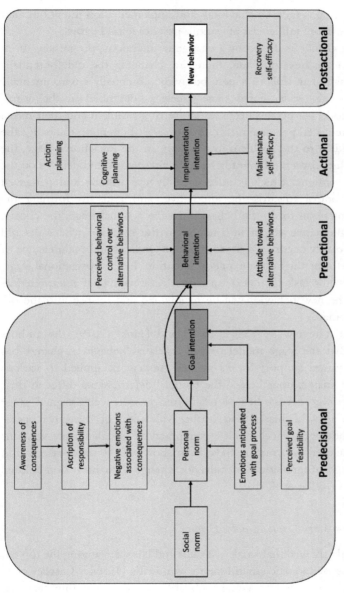

Fig. 2: The stage model of self-regulated behavioral change. (Source: Adapted from Journal of Environmental Psychology, 33, Bamberg, Applying the stage model of self-regulated behavioral change in a car use reduction intervention, p. 69, Copyright 2013 by Elsevier)

The four action phases (predecisional, preaction, actional and postactional) —understood as stages of decision-making—build the frame whereas the different intentions (goal-, behavioral- and implementation intention) alter the stage transition, influenced by various psychological factors.

Behavior change is driven by a change in intentions, progressing from one stage to the next one and, therefore, changing the old (potentially environmentally harmful) to a new behavior (potentially environmentally friendly). In the predecisional stage—such as described in the norm-activation model—a person becomes aware of a problem and their moral obligation (activated personal norm) to form a goal intention to change the behavior leading to the problem in question. In the preactional stage, the person weighs between different behavioral alternatives and decides on the most fitting options, which is influenced by the attitude and perceived behavioral control towards the alternative, and leads to forming the behavioral intention to do the behavior. In the actional stage, the chosen behavioral alternative is put into action in the situation where the old behavior normally occurs. To do so, planning abilities and maintenance of self-efficacy forms the implementation intention. In the postactional stage, it is the person's task to reflect on their decisions and, if interpreted as successful, the new behavior will be maintained and relapse into the old behavior avoided.

In a recent review by Keller, Eisen and Hanss (2019), the authors concluded that the stage model of self-regulated behavioral change has received empirical support for its general framework applied to various behavioral domains; nonetheless, the models' determinants differ in their prediction power across behaviors. Moreover, the self-assigned stage membership may be biased and further validation of measures operationalizing the stages is needed. However, practically speaking, interventions with stage-tailored information have been shown to be more effective in fostering stage progression and behavior change than have non-tailored interventions (Keller et al. 2019).

Example: From Single-Use to Reusable Drink Cups

A recent study showed that single-use cups and lids were among the top ten items littering our aquatic environments worldwide (Morales-Caselles et al. 2021). Paper cups tend to come with an internal plastic film that can discharge microplastics when used with hot water (Ranjan et al. 2021).

Consumer reduction of single-use cups thus is beneficial for the natural environment and potentially for health. Economic incentives alone do not seem to trigger the needed reduction in single-use cups and a mix of measures, including internal motivators, are needed (Poortinga and Whitaker 2018; Sandhu et al. 2021). Therefore, underlying psychological influences of alternative sustainable behaviors such as using a refundable cup from a city-wide deposit scheme, bringing one's own cup and reducing one's consumption to tackle the problem have been investigated (Keller et al. 2021).

The study by Keller et al. (2021) used the stage model of self-regulated behavioral change as the underlying framework and found that it partly explained single-use cup consumption and its sustainable alternative behaviors. For example, stronger implementation intentions predicted (self-reported) reduced consumption and own cup usage. That means, people reduced their consumption when they were outdoors or used their own cup to buy a take-away drink when they actively planned their action in the specific situation beforehand.

Keller et al. (2021) suggest that interactive campaigns such as websites or apps can be used to check people's stage and follow up with stage-appropriate information (such as in Klöckner and Ofstad 2017) or help them find the needed information (such as in Sunio et al. 2018). For example, Klöckner and Ofstad (2017) designed a website with stage-appropriate information in three different categories: 1) Why should I do something?; 2) What can I do?; How do I master the challenge? Each section told the story of three people and their individual goals to reduce their consumption (in this case beef). For example, the first subpage had information about why they reduced their consumption (targeting the preaction stage). The second subpage had information about how they reduced their consumption (targeting the preaction stage) and the third subpage had information about their challenges and how they overcame them (targeting the action and preaction stage).

Moreover, Keller et al. (2021) approve of further regulatory measures and propose that labelling products as non-recyclable could be beneficial to enhance the problem awareness of consumers. Moreover, labelling products clearly as non-recyclable, linked with a picture of the harm they can cause (e.g., a picture of a local animal eating microplastics), could increase the feeling of moral obligation even stronger.

> **Box 2: Behavioral Barriers—The Dragons of Inaction**
>
> Plastic reduction behavior such as shopping plastic-free can be demanding and impractical for consumers. People may encounter structural or situational barriers regarding their purchase behaviors e.g. encountering long ways to the zero-waste supermarket or farmers market, financial constraints when buying more expensive unpacked alternatives, no availability of unpacked alternatives in the local supermarket etc., which hinder the reduction of plastic waste even when consumers are motivated. However, researchers investigating environmentally friendly behaviors—including behaviors such as buying green and recycling—have also emphasized the importance of psychological barriers even where there are no structural or situational barriers (Gifford 2011; Gifford et al. 2018; Gifford and Chen 2017; Lacroix et al. 2019). Knowing these barriers and investigating which ones are hindering the implementation of the sustainable action can help practitioners to understand their target population and therefore, design tailored measures to foster sustainable change (Lacroix et al. 2019).
>
> One popular concept focusing on the psychological barriers of people is the so-called "Dragons of Inaction" (Gifford 2011). Lacroix, Gifford and Chen (2019) developed the *Dragons of Inaction Psychological Barriers Scale* measuring five barrier domains (change unnecessary, conflicting goals and aspiration, interpersonal relations, lacking knowledge, and tokenism). *Change unnecessary* describes the person's general belief that their individual action is not needed along with a denial of the importance to act and problems concerning the natural environment. *Conflicting goals and aspirations* represent perceived barriers such as strong habits, time constraints and fear of failure. *Interpersonal relations* cover feelings such as embarrassment and worry because of social disapproval. *Lacking knowledge* is based on a person's confusion about environmental topics and their uncertainty of where to get information from—also described by the authors as "a person's claim of ignorance, that one simply does not know how to change" (Lacroix et al. 2019, 11). *Tokenism* consists of the person's beliefs that responsibility lies in external entities such as industry, as they cause bigger environmental damage than oneself and the belief that one's personal actions are "enough". Lacroix et al. (2019) propose to use the scale to investigate if interventions are successful in reducing the perceived barriers and with that enhancing sustainable behavior.

A similar approach was suggested by Pahl et al. (2017) to connect people's passion for the ocean to daily behaviors and foster a likely reduction of individual plastic product consumption and their proper disposal. Following up on that, Luo et al. (2021) showed that visualizing marine consequences of plastics at recycling bins in an office building in Canada decreased plastic waste by 17 percent.

Interventions: What Needs to be Considered for Successful Behavioral Interventions?

A recent scoping review on behavior interventions focusing on plastic pollution concluded that research and assessment methods need to be improved as approaches differ greatly. Moreover, the review stated that measurement of actual behaviors (instead of predictors) and their impacts on the plastic systems are currently lacking (Nuojua et al. 2021). Therefore, having covered the recent theoretical background quite extensively, we will now provide some practical guidelines on how to implement behavior-targeted interventions. The *four key issues for encouraging pro-environmental behavior* and the *community-based social marketing approach* will be presented, followed by a brief introduction of useful intervention tools such as cognitive dissonance, goal setting, social modelling, and prompts. We do not want to reinvent the wheel, and we cannot present all approaches comprehensively. Instead, we selected established concepts that are hopefully useful to a wider interdisciplinary audience, encouraging collaboration to tackle plastic pollution beyond one single discipline and enhance the understanding of the importance of behavioral approaches.

Implementation of Interventions

Successful and effective behavior-targeted interventions need to be developed and evaluated systematically (McKenzie-Mohr and Schultz 2014; Steg and Vlek 2009). Therefore, social psychologists have developed a set of guidelines to ensure high-quality interventions. Both approaches are focusing on the individual as the change agent within the community.

The *four key issues for encouraging pro-environmental behavior* (Steg and Vlek 2009) are divided in the identification of the behavior in need of change (1) and behavior relevant determinants (2) mentioned earlier, followed by designing the interventions with the appropriate strategies, its application (3) and evaluation (4). Below we display the leading questions for the systematic process of interventions proposed by Steg and Vlek (2009, 310):

I. Which behaviors should be changed to improve environmental quality?
1. Select behaviors having significant negative environmental impacts
2. Assess the feasibility of behavior changes
3. Assess baseline levels of target behaviors
4. Identify groups to be targeted

II. Which factors determine the relevant behavior?
1. Perceived costs and benefits
2. Moral and normative concerns
3. Affect
4. Contextual factors
5. Habits

III. Which interventions could best be applied to encourage pro-environmental behavior?
1. Informational strategies (information, persuasion, social support and role models, public participation)
2. Structural strategies (availability of products and services, legal regulation, financial strategies)

IV. What are the effects of interventions?
1. Changes in behavioral determinants
2. Changes in behaviors
3. Changes in environmental quality
4. Changes in individuals' quality of life

A similar approach, proposed as guidance for practitioners, emerged a few years later named *Community-based social marketing* (McKenzie-Mohr 2011; McKenzie-Mohr and Schultz 2014). This framework consists of several steps. The first three steps are similar to the approach proposed by Steg and Vlek (2009). Step four and five of the community-based social marketing framework consider the context of the intervention's implementation more carefully than the approach above. The approach is broadly centered around 1) the thorough choice of which behavior(s) to target; 2) the detection of the associated barriers and benefits; 3) the development of strategies based on effective tools suitable for addressing the barriers and benefits; 4) the trialing of the intervention with a small sample and once shown as effective its (5) broad implementation and ideally (mid-and long-term) evaluation (McKenzie-Mohr 2011; McKenzie-Mohr and Schultz 2014).

Following these guidelines should result in better, hence, effective interventions as they are based on "carefully studied" tools (McKenzie-Mohr and Schultz 2014). Nevertheless, the intervention tools in the next section are context and behavior dependent and therefore, are unlikely to work in all contexts and for all target groups. It is important to systematically analyze why interventions did and why they did not result in behavior change as well as sharing these results as this will help to improve the development of interventions. Moreover, it is crucial to acknowledge that the community-based social marketing approach is used by researchers as well as by practitioners. This is a strength as it is versatile, but can also be a weakness as it can lead to misuse if it is not applied correctly. Therefore, intervention effectiveness can vary greatly. For instance, interventions in an intercultural setting which proofed to be effective in one community might not work for another due to the differences in personal, cultural or situational factors (Bosse 2010; Leenen 2005), especially when the intervention disrupts peoples every-day life (Richter, Grünzner and Klöckner, under review) such as using a playful tool like a board game as intervention to foster discussions about difficult and even conflict evoking topics (e.g., http://www.savannalife.no/).

For example, a study with Norwegian students and university employees found that proper waste disposal at work was mainly motivated by intentions, perceived behavioral control, personal and social norms as well as habits (Ofstad et al. 2017). Whereas, a study with an island community in Indonesia found that individual factors such as awareness were not considered when they thought about plastic pollution. For them, social-factors such as collective beach cleans or waste disposal organized by the community were things which came to mind first (Phelan et al. 2020).

In the first and second step of the community-based social marketing approach, it is important to listen to what the community members—respectively members of the target group—have to say about their personal, cultural and situational setting. Moreover, listening to their experiences during and after the intervention is essential to not undermine their voices, to choose the fitting intervention and to evaluate it appropriately.

Intervention Tools

Various tools, grouped in informational (motivational-focused) and structural (context-focused) strategies can be used to foster environmental behavior change (Steg and Vlek 2009). Both strategies are important and their effectiveness is dependent on the target behavior. Structural strategies—also known as hard measures—aim to make sustainable action easier and harmful action harder in a specific context. This can take the form of physical changes (increasing proximity to recycling bins), but also the implementation of legal measures (banning primary microplastics in personal care products) fall into this category (Steg and Vlek 2009). They are especially useful when external barriers make sustainable behavior difficult and can potentially influence behavioral determinants indirectly. Informational strategies—also so-called soft measures —on the other hand, can help to increase public support for structural changes (Steg and Vlek 2009). They are especially beneficial when external barriers are low and the sustainable behavior in question is easy to implement for the individual (e.g., no perceived barriers present) (Steg and Vlek 2009).

The intervention tools which are going to be presented are based on the findings of a meta-analysis investigating the effectiveness of intervention-based research targeting different sustainable actions such as various kinds of recycling (public, curbside, central) and conservation behaviors (energy, water, gasoline etc.) in an experimental setting (Osbaldiston and Schott 2012).

We are going to present a short description of the treatments, which were explored in the meta-analysis—sorted from low to high engagement for the participants. *Making it easy* describes the reduction of barriers to make the more sustainable behavior the easier option such as reducing the distance of recycling bins (McKenzie-Mohr and Schultz 2014; Osbaldiston and Schott 2012). *Prompts* are simple cues closely presented to the behavior in question such as "put plastic in the yellow bin" (McKenzie-Mohr and Schultz 2014; Osbaldiston and Schott 2012). They need to be self-evident, obvious and should focus on behaviors that encourage a sustainable action instead of avoiding a non-sustainable one (McKenzie-Mohr and Schultz 2014).

Justifications (declarative information) are pieces of information that explain why a certain behaviour should be done e.g., by illustrating the pathway of waste and explaining how misplacement of recyclable items lead to landfills or contamination of the natural environment etc. (Osbaldiston and Schott 2012). *Instructions (procedural information)* are pieces of information, which explain how to do a certain behavior e.g., by explaining the proper disposal of different kinds of plastics (Osbaldiston and Schott 2012). *Rewards* are one kind of incentive and generally consist of a positive outcome initiated by a specific behavior (McKenzie-Mohr and Schultz 2014). However, in the meta-analysis, they were described as a "monetary gain that people received as a result of participating in the experiment" (Osbaldiston and Schott 2012, 272).[2] *Social modeling* describes a variety of tools—such as social diffusion or norms—in which information is passed on by someone who encourages a specific sustainable action (Osbaldiston and Schott 2012). *Social diffusion* describes the adoption of sustainable behaviors by people because of the influence of significant others who already act sustainably. *Normative messages* can take form in information about how other people act and keeping the referent group more generic or self-determined is advised (McKenzie-Mohr and Schultz 2014). *Cognitive dissonance*—e.g., the conflict between the underlying pro-environmental attitude and environmental harmful behavior—was achieved by techniques such as foot-in-the-door. Meaning that participants were invited to do a smaller task—one to which participants easily agree—followed by an invitation to a more extensive one which is the actual targeted behavior (Osbaldiston and Schott 2012). *Feedback* is described as information that was given about past behavior over a certain amount of time (Osbaldiston and Schott 2012). *Commitment* to carry out a certain behavior was implemented in the interventions in verbal or written form e.g., signing a pledge card (Osbaldiston and Schott 2012). *Goal setting* in the interventions was advocated by providing the participants with a fixed goal, provided by the researchers, such as reducing their consumption by a specific amount in a certain time period (McKenzie-Mohr and Schultz 2014).

The researchers found that interventions in which cognitive dissonance, goal setting, social modeling or prompts were present were the most effective. However, intervention success across various behaviors differed and the authors concluded that "low-engagement treatments are appropriate

---

[2] The context in which the reward is placed is crucial when interpreting the results. It needs to be distinguished if participants are rewarded for taking part in the experiment itself or if they are rewarded when showing the sustainable behavior during the experiment.

for low-effort behaviors and high-engagement treatments are effective for high-effort behaviors" (Osbaldiston and Schott 2012, 280). Moreover, most interventions had multiple treatments in place. The most successful combinations were: Instructions and goals, prompts and making it easy, cognitive dissonance and justification, prompts and justification, rewards and goals, commitment and goals. Feedback and instructions seemed to have a smaller effect in comparison to the other treatments listed below. Moreover, feedback did not appear to be as effective in combination with other treatments in comparison to the presented alternatives.

Overall, each of the treatments displayed has been implemented in the interventions in various ways as main treatment or support. The meta-analysis summarized various treatments, but there are far more techniques at hand which have not been studied thoroughly. Nevertheless, the researchers were able to quantify the effectiveness of treatments across different environmental behavior domains. Recycling is one of the most extensively studied behaviors and even though it plays an important role in the reduction of plastic pollution, it is also important to look into consumption behaviors across different plastic sources to reduce the use of plastic overall. Therefore, more experiments looking into behaviors causing plastic and microplastics pollution is needed. Moreover, a systematic review focusing on the effectiveness of various intervention techniques or—if enough studies are found—a meta-analysis to quantify the various effects can help to create extensive research informed advice for practitioners.

Conclusion

In this chapter, we gave insights into environmental psychology approaches and their potential to increase the effectiveness of environmental behavior change interventions targeting plastic pollution.

To summarize, there is no one fits all solution when wanting to tackle environmental behavior. Some behaviors have stronger moral components, some are more influenced by the individual's perception of control (how easy or difficult it is doing a behavior) and others are strongly habitualized. It is key to understand these characteristics of the target behavior and also the system in which it takes place: Will a change of the behavior lead to an effective outcome in comparison to alternative measures? Will it be feasible to change considering internal and external constraints? If so, further development of the intervention can follow, using the series of steps introduced above, under consideration of the target-specific behavioral drivers and barriers.

Finally, behavior change is one part of the solution to reduce plastics in our natural environment, as "the consumer" plays one role in a complex system. A reduction in consumption behavior and an increase in recycling behavior is a start and can empower individuals. Nevertheless, it will not solve the problem on its own. Harmonized and immediate actions from different stakeholders such as governments, businesses, and communities worldwide combining pre- and postconsumption solutions are needed to achieve the necessary reduction of plastic polluting on our planet (Lau et al. 2020). For that reason, governments expressed their willingness to sign an international agreement tackling plastic pollution during the last meeting of the United Nations Environment Assembly and researchers are currently pushing for an "international legally binding agreement" targeting plastics on its complete life cycle (Simon et al. 2021).

Acknowledgements

We thank Isabel Richter and Mathew White for their helpful comments on an earlier version of this chapter.

This work has received funding from the European Union's Horizon 2020 research and innovation program under Marie Skłodowska-Curie grant agreement No 860720. The responsibility for the content in this chapter lies entirely with the authors.

A preprint of this chapter was submitted as a deliverable for the project LimnoPlast—Microplastics in Europe's Freshwater Ecosystems: From Sources to Solutions.

Author contributions

Maja Grünzner conceptualized the outline and wrote the manuscript.
Sabine Pahl supervised and edited the work.

References

Adane, Legesse, and Diriba Muleta (2011). Survey on the usage of plastic bags, their disposal and adverse impacts on environment: A case study in Jimma City, Southwestern Ethiopia. *Journal of Toxicology and Environmental Health Sciences*, 3 (8), 234–248.

Ajzen, Icek (1991). The theory of planned behavior. *Organizational Behavior and Human Decision Processes*, 50 (2), 179–211.

Bamberg, Sebastian (2013). Changing environmentally harmful behaviors: A stage model of self-regulated behavioral change. *Journal of Environmental Psychology*, 34, 151–159.

Bertoldo, Raquel, and Paula Castro (2016). The outer influence inside us: Exploring the relation between social and personal norms. *Resources, Conservation and Recycling*, 112, 45–53.

Bosse, Elke (2010). Vielfalt erkunden–ein Konzept für interkulturelles Training an Hochschulen. In Gundula Gwenn Hiller and Stefanie Vogler-Lipp (eds.). *Schlüsselqualifikation Interkulturelle Kompetenz an Hochschulen*, 109–133. Wiesbaden: VS Verlag für Sozialwissenschaften.

Braun, Yvonne A., and Assitan Sylla Traore (2015). Plastic bags, pollution, and identity: Women and the gendering of globalization and environmental responsibility in Mali. *Gender and Society*, 29 (6), 863–887.

Davison, Sophie, Mathew P. White, Sabine Pahl, Tim Taylor, Kelly Fielding, Bethany R. Roberts, Theo Economou, Oonagh McMeel, Paula Kellett, and Lora E. Fleming (2021). Public concern about, and desire for research into, the human health effects of marine plastic pollution: Results from a 15-country survey across Europe and Australia. *Global Environmental Change*, 69, 102309.

Dunlap, Riley., Kent Van Liere, Angela Mertig, and Robert Emmet Jones (2000). Measuring endorsement of the new ecological paradigm: A revised NEP scale. *Journal of Social Issues*, 56 (3), 425–442.

European Commission (2020). Attitudes of European citizens towards the environment (No. 501; Special Eurobarometer). 15.07.2021 https://europa.eu/eurobarometer/surveys/detail/2257

Gifford, Robert (2011). The dragons of inaction: Psychological barriers that limit climate change mitigation and adaptation. *American Psychologist*, 66 (4), 290.

Gifford, Robert, and Angel K. S. Chen (2017). Why aren't we taking action? Psychological barriers to climate-positive food choices. *Climatic Change*, 140 (2), 165–178.

Gifford, Robert, Karine Lacroix, and Angel K. S. Chen (2018). 7—Understanding responses to climate change: Psychological barriers to mitigation and a new theory of behavioral choice. In Susan Clayton and Christie Manning (eds.). *Psychology and climate change. Human perceptions, impacts, and responses*, 161–183. Cambridge: Academic Press.

Gollwitzer, Peter (1990). Action phases and mind-sets. In E. Tory Higgins and Richard M. Sorrentino (eds.). *Handbook of motivation and cognition: Foundations of social behavior*, 2, 53–92. New York: The Guilford Press.

Grünzner, Maja, Sabine Pahl, Mathew White, and Richard C. Thompson (2021). *Expert perceptions about microplastics pollution: Potential sources and solutions*. Study presented online at the Society for Risk Analysis – Europe conference, 13–16 June 21, in Espoo, Finland.

Felipe-Rodriguez, Marcos, Gisela Böhm, and Rouven Doran (2022). Risk perception: The case of microplastics. A discussion of environmental risk perception focused on the microplastic issue. In Johanna Kramm and Carolin Völker (eds.). *Living in the plastic age. Perspectives from humanities, social sciences and environmental sciences*. Frankfurt, New York: Campus.

Heidbreder, Lea Marie, Isabella Bablok, Stefan Drews, and Claudia Menzel (2019). Tackling the plastic problem: A review on perceptions, behaviors, and interventions. *Science of the Total Environment*, 668, 1077–1093.

Joanes, Tina, Wencke Gwozdz, and Christian Klöckner (2020). Reducing personal clothing consumption: A cross-cultural validation of the comprehensive action determination model. *Journal of Environmental Psychology*, 71, 101396.

Keller, Anna, Charis Eisen, and Daniel Hanss (2019). Lessons learned from applications of the stage model of self-regulated behavioral change: A review. *Frontiers in Psychology*, 10, 1091.

Keller, Anna, Jana Katharina Köhler, Charis Eisen, Silke Kleihauer, and Daniel Hanss (2021). Why consumers shift from single-use to reusable drink cups: An empirical application of the stage model of self-regulated behavioural change. *Sustainable Production and Consumption*, 27, 1672–1687.

Klöckner, Christian (2013a). A comprehensive model of the psychology of environmental behaviour—A meta-analysis. *Global Environmental Change*, 23 (5), 1028–1038.

Klöckner, Christian (2013b). How powerful are moral motivations in environmental protection?: An integrated model framework. In Karin Heinrichs, Fritz Oser, and Terence Lovat (eds.). *Handbook of moral motivation*, 447–472. Rotterdam: SensePublishers.

Klöckner, Christian (2015). *The psychology of pro-environmental communication: Beyond standard information strategies*. London: Palgrave Macmillan.

Klöckner, Christian, and Anke Blöbaum (2010). A comprehensive action determination model: Toward a broader understanding of ecological behaviour using the example of travel mode choice. *Journal of Environmental Psychology*, 30 (4), 574–586.

Klöckner, Christian, and Sunita Prugsamatz Ofstad (2017). Tailored information helps people progress towards reducing their beef consumption. *Journal of Environmental Psychology*, 50, 24–36.

Lacroix, Karine, Robert Gifford, and Angel Chen (2019). Developing and validating the Dragons of Inaction Psychological Barriers (DIPB) scale. *Journal of Environmental Psychology*, 63, 9–18.

Lau, Winnie, Yonathan Shiran, Richard Bailey, Ed Cook, Martin Stuchtey, Julia Koskella, Costas Velis, Linda Godfrey, Julien Boucher, Margaret Murphy, Richard Thompson, Emilia Jankowska, Arturo Castillo Castillo, Toby Pilditch, Ben Dixon, Laura Koerselman, Edward Kosior, Enzo Favoino, Jutta Gutberlet, Sarah Baulch, Meera Atreya, David Fischer, Kevin He, Milan Petit, Rashid Sumaila, Emily Neil, Mark Bernhofen, Keith Lawrence, and James Palardy (2020). Evaluating scenarios toward zero plastic pollution. *Science*, 369 (6510), 1455–1461.

Leenen, Wolf-Rainer (2005). Interkulturelle Kompetenz: Theoretische Grundlagen. In Leenen, Wolf-Rainer, Harald Grosch, and Andreas Groß (eds.). *Bausteine zur interkulturellen Qualifizierung der Polizei*, 63-110. Münster: Waxmann.

Luís, Sílvia, Catarina Roseta-Palma, Marta Osório de Matos, Maria Lima, and Cátia Sousa (2020). Psychosocial and economic impacts of a charge in lightweight plastic carrier bags in Portugal: Keep calm and carry on? *Resources, Conservation and Recycling*, 161, 104962.

Luo, Yu, Jeremy Douglas, Sabine Pahl, and Jiaying Zhao (2022). Reducing plastic waste by visualizing marine consequences. *Environment and Behavior*, 54 (4), 809–832.

Martinho, Graça, Natacha Balaia, and Ana Pires (2017). The Portuguese plastic carrier bag tax: The effects on consumers' behavior. *Waste Management*, 61, 3–12.

McKenzie-Mohr, Doug (2011). *Fostering sustainable behavior: An introduction to community-based social marketing*. Gabriola Island: New society publishers.

McKenzie-Mohr, Doug, and Paul Wesley Schultz (2014). Choosing effective behavior change tools. *Social Marketing Quarterly*, 20 (1), 35–46.

Morales-Caselles, Carmen, Josué Viejo, Elisa Martí, Daniel González-Fernández, Hannah Pragnell-Raasch, Ignacio González-Gordillo, Enrique Montero, Gonzalo Arroyo, Georg Hanke, Vanessa Salvo, Oihane Basurko, Nicholas Mallos, Laurent Lebreton, Fidel Echevarría, Tim van Emmerik, Carlos Duarte, José Gálvez, Erik van Sebille, Francois Galgani, Carlos García, Peter Ross, Ana Bartual, Christos Ioakeimidis, Groka Markalain, Atsuhiko Isobe, and Andres Cozar (2021). An inshore–offshore sorting system revealed from global classification of ocean litter. *Nature Sustainability*, 4 (6), 484–493.

Nielsen, Kristian Steensen (2017). From prediction to process: A self-regulation account of environmental behavior change. *Journal of Environmental Psychology*, 51, 189–198.

Nielsen, Kristian Steensen, Viktoria Cologna, Florian Lange, Cameron Brick, and Paul Stern (2021). The case for impact-focused environmental psychology. *Journal of Environmental Psychology*, 74, 101559.

Nielsen, Kristian Steensen, Theresa Marteau, Jan Bauer, Richard Bradbury, Steven Broad, Gayle Burgess, Mark Burgman, Hilary Byerly, Susan Clayton, Dulce

Espelosin, Paul Ferraro, Brendan Fisher, Emma Garnett, Julia Jones, Mark Otieno, Stephen Polasky, Taylor H. Ricketts, Rosie Trevelyan, Sander van der Linden, and Diogo Veríssimo (2021). Biodiversity conservation as a promising frontier for behavioural science. *Nature Human Behaviour*, 5 (5), 550–556.

Niinimäki, Kirsi, Greg Peters, Helena Dahlbo, Patsy Perry, Timo Rissanen, and Alison Gwilt (2020). The environmental price of fast fashion. *Nature Reviews Earth and Environment*, 1 (4), 189–200.

O'Brien, Joshua, and Gladman Thondhlana (2019). Plastic bag use in South Africa: Perceptions, practices and potential intervention strategies. *Waste Management*, 84, 320–328.

Ofstad, Sunita Prugsamatz, Monika Tobolova, Alim Nayum, and Christian Klöckner (2017). Understanding the mechanisms behind changing people's recycling behavior at work by applying a comprehensive action determination model. *Sustainability*, 9 (2), 204.

Ohtomo, Shoji, and Susumu Ohnuma (2014). Psychological interventional approach for reduce resource consumption: Reducing plastic bag usage at supermarkets. *Resources, Conservation and Recycling*, 84, 57–65.

Osbaldiston, Richard, and John Paul Schott (2012). Environmental sustainability and behavioral science: Meta-analysis of proenvironmental behavior experiments. *Environment and Behavior*, 44 (2), 257–299.

Pahl, Sabine, Isabel Richter, and Kayleigh Wyles (2022). Human Perceptions and Behaviour Determine Aquatic Plastic Pollution. In Friederike Stock, Georg Reifferscheid, Nicole Brennholt, and Evgeniia Kostianaia (eds.). *Plastics in the aquatic environment – Part II. Stakeholder's role against pollution*. Cham: Springer.

Pahl, Sabine, and Wyles, Kayleigh (2017). The human dimension: How social and behavioural research methods can help address microplastics in the environment. *Analytical Methods*, 9 (9), 1404–1411.

Pahl, Sabine, Kayleigh Wyles, and Richard Thompson (2017). Channelling passion for the ocean towards plastic pollution. *Nature Human Behaviour*, 1 (10), 697–699.

Phelan, Anna, Helen Ross, Novie Andri Setianto, Kelly Fielding, and Lengga Pradipta (2020). Ocean plastic crisis—Mental models of plastic pollution from remote Indonesian coastal communities. *PloS One*, 15 (7), e0236149.

Poortinga, Wouter, and Louise Whitaker (2018). Promoting the use of reusable coffee cups through environmental messaging, the provision of alternatives and financial incentives. *Sustainability*, 10, 873.

Ranjan, Ved Prakash, Anuja Joseph, and Sudha Goel (2021). Microplastics and other harmful substances released from disposable paper cups into hot water. *Journal of Hazardous Materials*, 404, 124118.

Richter, Isabel, Maja Grünzner and Christian A. Klöckner (2022). Global disruptive communication: The thin line between destruction and disruption in intercultural research. In Christian A. Klöckner and Erica Löfström (eds.). *Disruptive environmental communication – Shaking the tree, shaking humanity* [Unpublished manuscript].

Romero, Cláudia Buhamra Abreu, Michel Laroche, Golam Mohammed Aurup, and Sofia Batista Ferraz (2018). Ethnicity and acculturation of environmental attitudes and behaviors: A cross-cultural study with Brazilians in Canada. *Journal of Business Research*, 82, 300–309.

Sandhu, Sukhbir, Sumit Lodhia, Alana Potts, and Robert Crocker (2021). Environment friendly takeaway coffee cup use: Individual and institutional enablers and barriers. *Journal of Cleaner Production*, 291, 125271.

SAPEA, Science Advice for Policy by European Academies. (2019). A scientific perspective on microplastics in nature and society. Berlin: SAPEA.

Schultz, P. Wesley (2011). Conservation means behavior. *Conservation Biology*, 25 (6), 1080–1083.

Schwartz, Shalom H. (1977). Normative influences on altruism. *Advances in Experimental Social Psychology*, 10, 221–279.

Schwartz, Shalom H., and Judith A. Howard (1981). A normative decision-making model of altruism. *Altruism and Helping Behavior*, 189–211.

Schwarzer, Ralf (2008). Modeling health behavior change: How to predict and modify the adoption and maintenance of health behaviors. *Applied Psychology*, 57 (1), 1–29.

Sheeran, Paschal, and Thomas L. Webb (2016). The intention–behavior gap. *Social and Personality Psychology Compass*, 10 (9), 503–518.

Siegfried, Max, Albert A. Koelmans, Ellen Besseling, and Carolien Kroeze (2017). Export of microplastics from land to sea. A modelling approach. *Water Research*, 127, 249–257.

Simon, Nils, Karen Raubenheimer, Niko Urho, Sebastian Unger, David Azoulay, Trisia Farrelly, Joao Sousa, Harro van Asselt, Giulia Carlini, Christian Sekomo, Maro L., Per-Olof Busch, Nicole Wienrich, and Laura Weiand (2021). A binding global agreement to address the life cycle of plastics. *Science*, 373 (6550), 43–47.

Steg, Linda. (2018). Limiting climate change requires research on climate action. *Nature Climate Change*, 8 (9), 759–761.

Steg, Linda, Agnes E. van den Berg, and Judith I. M. de Groot (2018). Environmental psychology: History, scope, and methods. In Linda Steg and Judith I. M. de Groot (eds.). *Environmental psychology: An introduction*, 1–11. Hoboken, NJ: Wiley.

Steg, Linda, and Charles Vlek (2009). Encouraging pro-environmental behaviour: An integrative review and research agenda. *Environmental Psychology on the Move*, 29 (3), 309–317.

Stern, Paul C. (2000). Toward a coherent theory of environmentally significant behavior. *Journal of Social Issues*, 56 (3), 407–424.

Stern, Paul C., and Thomas Dietz. (1994). The value basis of environmental concern. *Journal of Social Issues*, 50 (3), 65–84.

Stieß, Immanuel, Luca Raschewski, and Georg Sunderer (2022). Everyday life with plastics: How to put environmental concern into practice(s). In Johanna Kramm and Carolin Völker (eds.). *Living in the plastic age. Perspectives from humanities, social sciences and environmental sciences.* Frankfurt, New York: Campus.

Sunio, Varsolo, Jan-Dirk Schmöcker, and Junghwa Kim (2018). Understanding the stages and pathways of travel behavior change induced by technology-based intervention among university students. *Transportation Research Part F: Traffic Psychology and Behaviour*, 59, 98–114.

van Valkengoed, Anne M., and Linda Steg (2019). Meta-analyses of factors motivating climate change adaptation behaviour. *Nature Climate Change*, 9 (2), 158–163.

Verplanken, Bas, and Henk Aarts (1999). Habit, attitude, and planned behaviour: Is habit an empty construct or an interesting case of goal-directed automaticity? *European Review of Social Psychology*, 10 (1), 101–134.

Whitmarsh, Lorraine, Wouter Poortinga, and Stuart Capstick (2021). Behaviour change to address climate change. *Current Opinion in Psychology*, 42, 76–81.

Chemicals in Plastic Packaging: Challenges for Regulation and the Circular Economy

Jane Muncke, Lisa Zimmermann

Plastics are complex chemical mixtures. They contain many different chemicals; some serving a specific function in the plastic product (such as additives that make plastic flexible or hard) and some having no functionality, such as impurities or degradation products. When addressing the issue of chemicals in plastics, it is reasonable to focus mostly on plastics used for food contact purposes, as here most data and information are available and these chemicals may directly transfer into foodstuffs where they are ingested by humans—although there is an increasing evidence base also for chemicals present in other, non-food contact plastics. We assume that plastic food contact materials (FCMs) are the most direct and relevant human exposure source to chemicals in plastics for the general population. Therefore, in this chapter, we zoom in on plastic and bioplastic FCMs. It is however important to note that plastic chemicals used for other applications than FCMs are also important. First, we discuss the chemical composition of food contact plastics, their risk assessment, regulation, and potential health consequences. Then, we evaluate the fate of plastics in a circular economy, including the impact of recycling, and the role of alternative materials such as bioplastics.

Food Contact Materials

The largest current use of plastics is for packaging (approx. 40 percent), with food and beverage packaging contributing the most to this industry sector: approximately 10–20 percent of the overall global plastics production is for food packaging. Plastic packaging that is intentionally brought into direct contact with food is also known as "food contact" plastic. Besides packaging, also other items contact food and are collectively called food contact articles (FCAs). FCAs cannot only consist of plastics but also of other materials like adhesives, printing inks, and paper referred to as food contact materials (FCMs) (see Fig. 1). Both FCAs and FCMs consist of and contain so-called food contact chemicals (FCCs) which is not a legal term but is still useful when discussing any and all the chemicals that are present in food packaging, regardless of whether they are intentionally added because they have a function, or not (Muncke et al. 2017). Importantly, plastics are not inert materials, meaning that they transfer some of their chemical constituents into surrounding media—for example into food and beverages. This transfer of chemicals from FCMs into foodstuffs is known as migration.

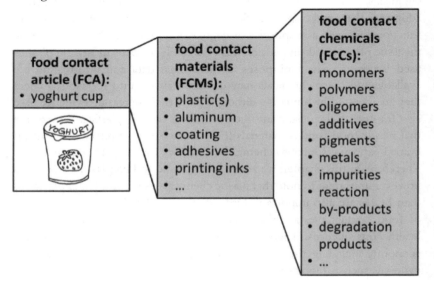

Fig. 1: Food contact articles (FCAs, e.g., yoghurt cup) are combinations of different food contact materials (FCMs, e.g., polypropylene and printing inks). FCAs and FCMs consist of and contain many food contact chemicals (FCCs) such as additives, un-reacted monomers and impurities. (Source: Muncke et al. 2017)

In today's societies, many of the foodstuffs that reach the consumer are packaged[1] (Costa et al. 2020). This packaging consists of different FCAs, made up of various FCMs (see Fig. 1), but the majority of food packaging and direct FCMs are plastic or synthetic, plastic-like materials such as the

---

1 In a 2020 Portuguese study, 57 percent of foodstuffs were packaged.

coatings on the inside of most food and beverage cans[2] (Poças et al. 2009; Costa et al. 2020). But food containers made of paperboard are also chemically treated, or even coated, with a thin plastic layer. In conclusion, it is important to understand that today's food packaging is highly engineered and consists of many different FCMs, and plastics are very common on today's market.

Tab. 1: Common types of food packaging and their food contact material constituents. Examples of migrating chemical types are given. (Source: Own draft)

Packaging type	Food contact materials	Chemical migration
Plastics	**Plastics**, adhesives, printing inks, paper, coatings	Monomers (styrene, caprolactam, bisphenol A); oligomers; plasticizers (ortho-phthalates, acetyl tributyl citrate, DEHA); catalysts (antimony); antioxidants (2,4-di-tert-butylphenol, Irgafos 168)
Paper & board	**Paper and cardboard**, printing inks, adhesives, plastics, metal	Printing inks (benzophenones, mineral oils, ortho-phthalates); grease-proofing (PFAS); antioxidants (butylated hydroxytoluene)
Metal	**Metal, coatings** and laquers, printing inks, paper and adhesives	Oligomers; monomers (BADGE, melamine, bisphenol A); slip agents (erucamide)
Glass	**Glass**, metal, plastic, coatings, printing inks, paper, adhesives, silicones, cork	Monomers (BADGE, bisphenol A); metal ions (lead, zinc, sodium, potalisium, barium)
Multilayer multi-material	**Plastic**, metal, **cardboard**, printing inks, adhesives	Printing inks, monomers

Chemical Composition of Plastics

Plastics as a food packaging material consist of many different FCCs (see Tab. 1), and this makes plastic FCMs highly complex. This chemical complexity is partially due to the way plastics are commonly manufactured: plastics' main, structure-giving constituents are polymers. Polymers are large, synthetic (or man-made) molecules made up of up to 10,000 repeating molecular building blocks. These building blocks are small molecules known as monomers. Polymers are made by an aggressive chemical reaction process

---

[2] In the same 2020 Portuguese study, the food packaging types were 69 percent plastics, 14 percent multi-layer multimaterial (with plastic as direct food contact material), plus an unknown level of coated metal (synthetic, plastic-like material) and plastic-coated paper.

called polymerization. During polymerization, many reaction by-products are formed, especially if the starting materials (i.e., monomers) are not of high purity. In this case, chemical impurities can also react during polymerization and form new chemicals which are then also present in the finished plastics (but which have no intended function).

Besides polymers and reaction by-products from polymerization, plastics contain unreacted monomers and impurities as well as intentionally added additives. Additives can make up 50 percent of the plastics by weight. They impart material properties to plastics, such as flexibility or stiffness, color, durability (protection of the polymers from degradation), and so forth. In conclusion, a finished plastic FCM is a complex material consisting of hundreds to thousands of different chemicals—some very large, like polymers, which give plastics their formability and moldability, and some very small, like some of the additives and non-intentionally added substances (NIAS). NIAS include impurities, reaction by-products, and break-down products of additives.

All these chemicals that are present in plastics can transfer (or migrate) into foodstuffs. Migration is a well-studied phenomenon that is backed by an abundance of scientific studies from the last 50 years. Several principles determine the extent of chemical migration into foodstuffs:

(1) Levels of migrating chemicals in packaging—higher levels of chemicals in FCM lead to higher levels migrating into foodstuffs
(2) Temperature—migration occurs faster and at higher levels when temperatures increase, meaning migration is higher into hot foodstuffs, or during storage at room temperature compared to cooling conditions (freezer)
(3) Storage time—migration occurs over time, and longer storage times imply higher migration
(4) Food chemistry—migration of fat-soluble, lipophilic chemicals is faster into fatty foods, while migration of water-soluble, polar/hydrophilic chemicals occurs faster into aqueous foodstuffs
(5) Packaging size—smaller size packaging has a proportionally larger surface area in contact with food, per volume of foodstuff, compared to larger size packaging; therefore, migration in smaller packaged portion sizes is proportionally higher than in larger portion sizes

It is important to note that migration can occur even if an FCA is not directly touching a foodstuff, but if the foodstuff is in its proximity. This is the case when volatile chemicals migrate, for example from printed food packaging into (fatty) solid foods.

There are many examples for migration (Tab. 1) and in some special cases, chemicals migrating from (plastic) food packaging can also interact with the food in chemical reactions which form hazardous chemicals. For example, when benzoic acid migrates from plastic packaging into foodstuffs containing ascorbic acid, the carcinogen benzene can be formed as a reaction product (van Poucke et al. 2008).

Understanding migration is critical for assessing the safety of FCMs and FCAs. In the European Union, the current regulatory definition of safety for FCMs is based on migration. Article 3 of the so-called FCM Framework Regulation (EU 1935/2004) specifies that

"Materials and articles, including active and intelligent materials and articles, shall be manufactured in compliance with good manufacturing practice so that, under normal or foreseeable conditions of use, they do not transfer their constituents to food in quantities which could:
 (a) endanger human health; [...]" (EU 2004, 7)

In the US, food contact materials are considered "safe" if there is

"reasonable certainty in the minds of competent scientists that the substance is not harmful under the intended conditions of use" (Food Additive Amendment 1958, 3).

Both regulatory definitions of safety require that any migrating chemicals shall be assessed for their risk to human health. But this is not always feasible for various reasons, including that some NIAS cannot be identified, or that the pure chemical substance is not commercially available and therefore cannot be quantified with certainty nor empirically assessed for its hazard properties. Therefore, one has to assume that there are many FCAs available on the market from which untested chemicals migrate into food. Whether these untested chemicals present a harm to human health is unknown, and essentially unknowable, using current approaches to chemical risk assessment.

Toxicity of Plastic Chemicals

Chemicals that migrate from FCAs into packaged foodstuffs are taken up by humans when they ingest the food or beverage. In the past decades, plastic chemicals have been detected in virtually every human tested. For instance, metabolites of the plastic monomer bisphenol A (BPA) and phthalate ester plasticizers widely used as plastic additives have been measured in more than 90 percent of the US population (Silva et al. 2004; Calafat et al. 2008; Zota et al. 2014). A study of synthetic chemicals in human umbilical cord blood found around 3,500 different chemicals, including several known to be used in plastics (Wang et al. 2021).

Health Consequences of Plastic Chemicals

Humans are exposed to plastic chemicals that they ingest on a daily basis, and for which this chronic exposure can cause negative health outcomes. Human biomonitoring studies are carried out to determine the types of chemicals humans are exposed to and their levels found in people, where urine and blood samples (or any other type of human samples) are analyzed for certain known FCCs. To estimate the human health consequences of a chemical, it is conventionally administered to a large group of rodents (e.g., rats) over a certain time period. Afterward, the animals are examined for overt signs of toxicity such as a reduced number of offspring compared to a control group that was not exposed to the chemical. By testing different concentrations, the highest concentration without any effect, as well as the lowest concentration leading to an effect, can be determined. To derive a safe exposure level for humans, assessment factors are applied that account for the differences between the tested animal and humans as well as for uncertainties of the testing procedure. Instead of performing toxicity studies in living organisms (*in vivo*), an alternative approach is to conduct them outside their normal biological context such as in isolated cells or microorganisms (*in vitro*).

More than a hundred studies have demonstrated health consequences of BPA including effects on reproduction, the mammary gland, the immune system, brain structure, neurobehavior, and metabolic endpoints. These effects can already occur at low doses (Vandenberg et al. 2019). Like BPA, evidence that phthalates influence the endocrine system, the nervous and the immune systems, and the functioning of multiple organs is widespread (Maffini et al. 2021; Wang and Qian 2021).

However, in addition to the few well-studied plastic monomers and additives, plastics contain and leach many other chemicals: More than 12,000 intentionally-added substances may be used in the manufacturing of FCMs in general and more than 4,000 chemicals have been associated with plastic packaging in particular. Among these are substances of concern. For these, scientific studies have demonstrated that they can have an negative impact on human health and/or the environment (Groh et al. 2019; Groh et al. 2021). For the majority of plastic chemicals, hazards have not even been analyzed, meaning hazard data are lacking. As discussed in the previous section on the chemical composition of plastics, as a result, many plastic products on the market, including food packaging, are not tested for their chemical hazards. Moreover, most of the chemicals contained in and migrating from plastics are unknown, making it impossible to quantify them and to assess their health consequences (Zimmermann et al. 2021).

One approach to identify hazard properties of chemicals with unknown identity is to test the whole mixture of chemicals that migrate or are extractable from plastics for their mixture toxicity. At the same time, this approach also takes into account effects related to the interaction of the different substances and reflects the actual composition of chemicals and overall toxicity the consumer is likely exposed to. *In vivo* and *in vitro* test methods using cell-based or whole-organism bioassays indicate that chemical mixtures originating from plastic FCMs can be toxic to humans and wildlife alike. They can result in cell death (cytotoxicity), damage of the genetic material (genotoxicity), and can interfere with the hormone system (endocrine disruption) (Wagner and Oehlmann 2009; Groh and Muncke 2017; Severin et al. 2017; Zimmermann et al. 2019; Pinter et al. 2020).

The Toxicity of Plastics Is Product-Specific

Especially critical is that the chemical composition, as well as the toxicity, are highly specific to the plastic product. *In vivo* and *in vitro* studies suggest that products made of certain polymer types, such as PVC and polyurethane (PUR), more frequently contain and leach toxic chemicals than other polymer types (Lithner et al. 2009; Zimmermann et al. 2019; Zimmermann et al. 2021). Potential reasons for this are that PVC and PUR require high quantities of additives and have been ranked most hazardous based on the monomers they are made of (Lithner et al. 2011). At the other end of the spectrum, some studies have reported PET to be free of toxic chemicals (Bach et al. 2013; Mertl et al. 2014). However, other studies found toxic chemicals contained in and migrating from PET products (Yang et al. 2011). Therefore, chemical composition and toxicity of a product cannot be predicted based on the polymer type alone. Instead, whether a plastic item contains toxic chemicals and whether they transfer into foodstuffs or not is largely dependent on the individual product, its chemical composition and manufacturing.

These differences between plastic products are not discernable upon optical inspection. For instance, two products made of the same polymer (e.g., polyethylene, PE) may contain a very different set of intentionally and non-intentionally added substances, also leading to different toxicity profiles (Zimmermann et al. 2019). The same applies to two products serving the same purpose and of identical appearance (e.g., two yoghurt cups). Thus, neither the polymer type nor application and appearance can serve as an indicator of a product's safety or toxicity. This prevents giving robust recommendations to industry, retail, and consumers on how to reduce the use and exposure to hazardous chemicals that migrate from plastic food packaging, other than an avoidance of chemically complex FCAs in general.

On a positive note, based on currently known hazard properties of concern, safer products are already on the market. These products do not contain chemicals with well-known hazard properties, such as chemicals influencing the hormone system, or, they may contain these chemicals but they will not transfer into the food. Thus, they are not available for human consumption (Zimmermann et al. 2019; Zimmermann et al. 2021). These products could be used as best-practice examples in the design of safer plastics. As a prerequisite, manufacturers would need to disclose the chemical composition and processing steps of their plastic products, which is, at present, not a requirement for food packaging.

Risk Assessment of FCCs—Current Practice and Options for Improvement

In chemical risk assessment, the central paradigm is that higher exposure levels equal higher risks. This is derived from the 16th-century scholar Paracelsus' Third Defense, where he wrote in 1564: "What is there that is not poison? All things are poison and nothing (is) without poison. Solely the dose determines that a thing is not a poison" (Deichmann et al. 1986, 212). Paracelsus' observation is very much reproducible, but in chemical risk assessment, it is interpreted to mean that low doses are of lower risk, while higher doses are of higher risk. This is indeed a logical fallacy, especially since there is empirical evidence demonstrating that in some cases, lower doses of chemicals are indeed of higher risk than higher doses. We discuss this issue in more detail in the section on non-monotonic dose-response below.

The risk assessment of chemicals, as it is practiced today by regulatory authorities and industry worldwide, assumes that health risks are higher if exposure levels are higher, and if the hazard properties of a given chemical are considered very severe. Therefore, for some chemicals with hazard properties of high concern no safe levels exist. For example, genotoxicants, which can directly damage DNA in various ways that may lead to cancer in humans, are considered highly hazardous chemicals, for which the risk is high even if the levels of the chemical (in food, etc.) are low. As consequence, genotoxicants are not allowed for use in the manufacture of food packaging, but there are some notable exemptions such as vinyl chloride (ATSDR 2016) or formaldehyde: For these genotoxic substances, it is assumed that they are chemically changed during manufacture so that no residues of the genotoxic substances remain present which could migrate into foodstuffs. However, this is not always the case and data are showing that migration from FCMs does indeed seem to occur (Castle et al. 1996; Ebner et al. 2020).

For other, non-genotoxic chemicals, thresholds for human exposure are set at levels below which exposures are assumed to have an acceptably low risk. In other words, health risks are expected to be larger if the level of a chemical is higher. Based on this, migration of chemicals from food packaging and other types of FCAs is considered acceptable, as long as certain levels are not exceeded. In recent years, this approach to chemical risk assessment has been criticized as not being founded in current scientific understanding. As consequence, it is not considered to be sufficiently protective of human health—an opinion, that also has been reflected in a 2016 report by the European Parliament (2016). The outlined examples indicate that chemical risk assessment as currently practiced for FCMs has several severe shortcomings, which are explained in more detail in the following.

Risk Assessment of Unknowns and "Non-Standard" Chemicals Is Impossible Using Currently Conventional Approaches

Many of the chemicals contained in FCMs are unknown, e.g., in the complex chemical synthesis many unpredictable reaction by-products are generated (Bradley and Coulier 2007). And as studies show, some of these unknown chemicals also migrate into foodstuffs during the normal, intended use of an FCA (Qian et al. 2018; Kato and Conte-Junior 2021; Zimmermann et al. 2021). But if a chemical is unknown, it cannot be measured for its level of migration into foods as chemical analysis requires availability of the pure chemical to calibrate the analytical equipment, and its toxicity cannot be assessed as this also implies that the pure chemical substance is available. Therefore, unknown chemicals essentially cannot be tested and, subsequently, cannot be assessed for their risk to harm human health using today's conventional chemical risk assessment approaches. This reality is currently being accepted, and some "work around" tools like the use of prioritization approaches have been developed and applied[3] (Koster et al. 2011; Koster et al. 2014; Pieke et al. 2018; Zimmermann et al. 2021). However, these approaches for the prioritization of chemicals in plastics do not result in fully risk-assessed chemicals with the same level of certainty that conventional chemical risk assessment requires. As consequence, the risk to human health from unknown chemicals migrating from food packaging essentially remains unknown.

For example, every plastic contains oligomers—these are short-chain polymers, small molecules that are a normal by-product of polymerization. Oligomers are usually small molecules and therefore relevant for human health risk assessment, as they can migrate and be taken up in the gastro-

---

[3] For example, the Threshold of Toxicological Concern (TTC) concept is used in combination with non-targeted chemical analysis or bioassays of overall migrate. However, TTC is not a scientific method, it allows for approximation and prioritization with inherently high levels of uncertainty (Bschir 2017).

intestinal tract. But oligomers are difficult to identify and usually not available as pure chemicals for further testing. Therefore, the knowledge about their toxicity is usually limited and a risk assessment of oligomers is, therefore, most often not possible.

The Dose Does Not Always Make the Poison

Current chemical risk assessment assumes that "the dose makes the poison" and that risk increases (and decreases) in a dose-dependent way. But scientific studies on chemicals which interact with the hormone system, that is also known as the endocrine system, have shown that dose-effect relationships of such chemicals are not always linear (Vandenberg et al. 2012; Gore et al. 2015). In practice, both natural hormones, as well as certain synthetic man-made chemicals, can have effects at low levels which are not seen at higher levels. Here, the dose-response relationship follows a U-shape or an inverted U-shape curve, which means that intermediate concentrations have a lower or higher effect than low and high concentrations of the same compound. This phenomenon has been repeatedly found for some chemicals, and these substances are known as endocrine-disrupting chemicals as their effects will interfere with the functioning the endocrine system. Current standard chemical risk assessment does not consider such non-monotonic dose responses but instead assumes that effects of high levels can be extrapolated in a linear way to assess effects at low levels. Therefore, human health is currently not adequately protected from exposures to hormone-hacking chemicals that migrate from food packaging into food at lower levels.

For example, the chemicals BPA and di-ethyl hexyl phthalate (DEHP) have both been shown to have non-monotonic dose responses for certain biological effects (Takano et al. 2006; Montévil et al. 2020).

Certain Types of FCCs Should Not Be Used

There is a consensus in chemical risk assessment that exposure to (man-made) genotoxicants (i.e., chemicals that alter and damage genetic material and/or nucleic acids such as RNA or DNA) must be as low as possible. As consequence, such substances shall not be used intentionally in FCMs unless explicitly authorized and that their migration into foodstuffs must be below acceptable thresholds. But this approach of setting thresholds for genotoxicants and also for non-genotoxic carcinogens (i.e., chemicals that can lead to cancer by non-genotoxic mechanisms) is applied to single substances in food packaging and other FCAs. It does not factor in that humans are simultaneously exposed to a multitude of genotoxic and non-genotoxic substances at low levels from other, often natural, sources. Consequently, the risks of the individual substances accumulate. Some risks could be avoided by not authorizing known genotoxicants for use in products such as food packaging. By contrast, the naturally present genotoxicants are far less manageable, such as acetaldehyde in ripening fruit. Therefore, to reduce human risk from exposure to all types of genotoxicants, known and avoidable sources, even if they are at individually low levels, should be eliminated. Such an approach, aimed at eliminating low-level exposures to genotoxicants in the majority of the population reduces the overall risk for cancer caused by these chemicals in the entire population (Rose 1981).

For example, styrene is a probable human carcinogen (IARC 2018) and a monomer used in the manufacture of polystyrene food packaging such as the packaging of dairy products or plastic cold cups. It is inevitable that small amounts of unreacted styrene will migrate from polystyrene into food, thereby exposing a large part of the human population to a probably carcinogenic chemical. Therefore, the use of polystyrene food packaging should be eliminated to reduce human genotoxicant exposure from this known avoidable source.

Mixture Toxicity—the Reality, Not the Exemption

According to current regulations, chemicals migrating from food packaging into food must be assessed individually. However, migration does not happen for one chemical at a time only. Plastic FCMs are highly complex mixtures of many different chemicals. Some of these chemicals can transfer, again as a chemical mixture, into foodstuffs being ingested with the food. This chemical mixture can exert health effects, even if the individual chemicals in the mixture are below their effect thresholds. But today's chemical risk assessment approach does not consider that mixtures of chemicals migrate and that these mixtures can have other effects than the sum of the individual chemicals. Instead, chemicals are assessed one by one (if at all) and safe exposure thresholds are set for individual, intentionally added chemicals only.

Mixtures are a reality for all plastics in direct food contact, with hundreds to thousands of chemicals found to migrate at the same time, many of which are unknown (Zimmermann et al. 2021).

The Timing of Chemical Exposure Influences the Risk

Along with the level of exposure to a chemical, and the mixture with other chemicals, the timing of the chemical exposure also influences the risk. Timing here means the point in a human's life when exposure to a manmade chemical takes place. For example, the same level of chemical may not affect a 27-year-old adult, but be detrimental to a developing fetus or newborn. Indeed, during development exposure to certain chemicals can induce long-term adverse health impacts, which only manifest later in (adult) life. The Developmental Origins of Health and Disease (DOHAD) hypothesis implies just this. Several studies have linked low-level prenatal exposures to man-made chemicals with diverse health effects such as allergies, neurological disorders, obesity and diabetes, and cancers. Currently, the timing of chemical exposure and its effects on different life-stages are not rigorously assessed when chemicals for use in FCMs are authorized.

For example, the synthetic chemicals BPA, 4,4-bisphenol F (BPF), and perfluorooctanoic acid (PFAS) were measured (together with other man-made chemicals) in pregnant women during their first trimester. Increased levels of these chemicals were correlated with a reduced intelligence quotient in the boys seven years later (Tanner et al. 2020).

Today's Chemical Exposure Can Affect Future Generations

In addition to directly affecting the developing fetus, chemical exposures during development or later in life may "imprint" themselves on the genetic material. Imprinting means that the genetic code sequence is left unchanged, but changes are inferred as to how genes are activated. These changes are found on the DNA or "on the gene," and hence are named epigenetic.

Indeed, many chronic diseases seem to have epigenetic mechanisms instead of coming from changes to the genetic code's sequence (Feil and Fraga 2012). But what is more, certain epigenetic changes are heritable and transferred across several generations, even long after a chemical exposure occurred (Breton et al. 2021). As consequence, future generations may be at risk for detrimental health effects due to exposures which happened long before their existence, in their great-grandparents (or even before). Knowledge of such mechanisms implies that there is a moral obligation for today's decision-makers to factor in the protection of future generations. This is another issue presently barely discussed in the chemical risk assessment of FCMs currently. Food contact chemicals that are known to affect the epigenome are BPA (Dolinoy et al. 2007) and DEHP (Robles-Matos et al. 2021).

Conclusion: Challenges for Chemical Risk Assessment of FCMs and the Way Forward

Science is constantly advancing what is known about chemicals and their impacts on human health. Yet chemical risk assessment for FCAs is currently stuck with a toxicological paradigm that is outdated and has been disproved by modern science. As a result, current chemical risk assessments practiced by regulators and industry are not sufficiently protective of public health (Muncke et al. 2020). A new approach is needed which systematically addresses all known shortcomings and which leads to a better protection of the general public from chemicals that migrate from food packaging and other FCAs (Muncke 2021). The opportunity for health prevention is large: although exposures to individual chemicals from food packaging may be small, almost the entire population is exposed. In addition, even yet unborn generations are affected because the effects can be transgenerationally inherited. If these detrimental low-level exposures were removed, many would benefit as more cases of avoidable disease would be prevented in the general public. And, in addition, so far unknown consequences of mixture toxicity and cumulative risks would also be mitigated.

Regulatory Requirements and Enforcement Realities—Food Contact Plastics in Europe

In Europe, the Food Contact Materials and Articles Framework Regulation (EU 1935/2004) established the requirement to assess that the levels of chemicals migrating into food do not harm human health. This requirement is further for food contact plastics (EU 10/2011 with its amendments) where the need for risk assessment of NIAS is stipulated (Art. 19).

But regulatory requirements need to be controlled, and regulation must be enforced to establish that plastics on the market meet the safety demands set out in the regulation. This is challenging for several reasons. In Europe, around 1,000 chemicals are explicitly authorized for use in food contact plastics. Experts estimate that an additional four to six chemicals are present in finished food contact plastics for each authorized substance, meaning that over 4,000 chemicals would be expected to be present in finished plastics articles. Empirical studies show that this number can range between a few hundred to over 10,000 different chemicals in a given plastic product (Wagner et al. 2013; Zimmermann et al. 2021). So, assessing the risk of each of these chemicals separately would be challenging and practically not feasible. Risk assessment and enforcement are not always possible even for the chemicals that are explicitly authorized for use in food contact plastics, for example, because methods for their chemical analysis are not publicly available (Simoneau 2015).

It is unknown whether industry is performing risk assessments for each and every of these substances. What is clear is that regulatory enforcement agencies are neither assessing all chemicals that are authorized for use in food contact plastics nor controlling whether industry is meeting its regulatory requirements—in part, because the required information is not shared with authorities (McCombie et al. 2016). Indeed, enforcement experts estimate that regular controls are carried out for less than 100 food contact chemicals (McCombie 2018).

Currently, the EU Commission is revising its food contact materials regulation and it is unclear if the plastics regulation will be affected. However, it is clear that with increasing demands for using recycled plastics in food packaging, the issue of chemical risk assessment will only become more complex. For this reason, revised regulations must factor in current scientific understanding and address challenges related to the circular economy (Muncke et al. 2020). Consequently, new approaches for assessing the risks of chemicals in plastics should be developed.

Moving Forward—Testing the Safety of Chemicals in Plastics

A new approach for assessing the safety of chemicals in plastics starts with redefining "safety." Instead of assuming that for any man-made, synthetic chemical there is an exposure level that is of low risk, the default assumption should be that only chemicals that have no relevant hazard properties have safe exposure levels. As consequence, all chemicals that are present in plastic food packaging need to be empirically tested for their hazard properties. The relevant hazard properties are tied to a chemical's propensity to cause adverse effects and disease (such as cancer, diabetes, allergy, infertility, etc.) in humans, or to negatively affect the environment (for example by being persistent and accumulating in the food chain). For this purpose, novel testing approaches need to be developed and embedded in a new regulatory framework for assessing the health risks of synthetic chemicals present in plastics (Muncke 2021).

"Bioplastics"—Shifting Problems

With the increased public awareness of the downsides of plastics, that they are being made of non-renewable resources and accompanied by inadequate waste management, demands for more environmentally friendly alternatives to plastic food packaging have been rising. Bioplastics are traded as a promising replacement. Only accounting for one percent (2.1 million tonnes) of the plastics produced today, the bioplastics market is expected to grow by 36 percent until 2025 (2.8 million tonnes) (PlasticsEurope 2019; European Bioplastics 2020). Bioplastics can be bio-based materials made from renewable resources (e.g., bio-based polyethylene (PE) and polyethylene terephthalate (PET)), biodegradable materials meant to degrade by biological activity (e.g., polybutylene succinate, PBS), or materials that are both bio-based and biodegradable (e.g., polylactic acid, PLA; Fig. 2) (Lambert and Wagner 2017). A universally agreed-upon definition of the term "bioplastics" is still absent. Thus, it is currently unclear whether plant-based materials (e.g., starch-, cellulose-, and bamboo-blends), made from natural polymers but often blended with synthetic materials, would also fall under that category.

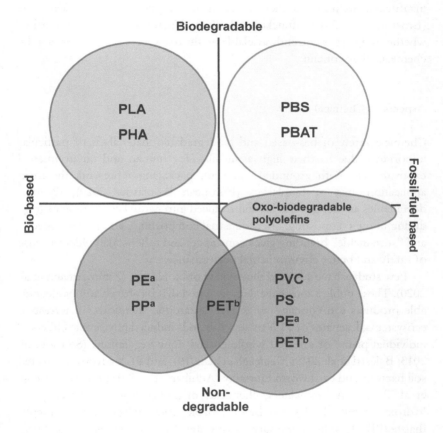

Fig. 2: Examples of biodegradable, bio- and fossil-fuel-based plastics. Note: [a]produced from bio-based or fossil-fuel-based resources. [b]usually made from fossil raw materials, but up to 30 percent bio-based starting materials can be included during production. Abbreviations: PLA, polylactic acid; PHA, polyhydroxyalkanoates; PBS, polybutylene succinate; PBAT, polybutylene adipate terephthalate; PE, polyethylene; PP, polypropylene; PVC, polyvinyl chloride; PS, polystyrene; PET, polyethylene terephthalate. (Source: adapted from Geueke 2014)

In comparison with petrochemical plastics, newly developed bio-based and biodegradable materials are commonly evaluated for performance and sustainability criteria in production (e.g., carbon dioxide footprint) and end-of-life (e.g., waste avoidance). However, chemicals currently receive insufficient attention in these assessments despite being a critical issue (Ernstoff et al. 2019; Muncke et al. 2020). Therefore, the question arises whether bioplastics currently available on the market are safe concerning the chemicals they contain.

Aspects of Chemical Safety

Chemical safety of bio-based and biodegradable materials is of particular importance due to their high availability for human and environmental exposure: (1) With around 47 percent, packaging represents the main application area of bioplastics (European Bioplastics 2020), (2) biodegradables are designed to be reintroduced into natural environments (e.g., during home composting, after industrial composting), and (3) the "green" and "sustainable" branding gives consumers and other stakeholders a sense of safety and better environmental performance.

Few studies have assessed the toxicity of bioplastics (Zimmermann et al. 2020). These publications have demonstrated that bio-based and biodegradable products can contain and release hazardous chemicals. In terrestrial ecosystems, leachates of PLA or starch-blends induced phytotoxic effects in individual plants or affected whole coastal dune vegetations (Souza et al. 2013; Balestri et al. 2019; Menicagli et al. 2019), and PLA adversely affected soil bacteria and earthworm mortality (Adhikari et al. 2016; Huerta-Lwanga et al. 2021). In freshwaters, chemicals migrating from bio-based polyhydroxybutyrate (PHB) and biodegradable polybutylene adipate terephthalate (PBAT) reduced the survival of water fleas (Göttermann et al. 2015). Studies related to human health aspects have shown that FCMs made of different bio-based and biodegradable materials often contain a high amount and wide variety of chemicals that can induce unspecific and endocrine effects *in vitro* (Maisanaba et al. 2014; Aznar et al. 2019; Asensio et al. 2020; Zimmermann et al. 2020). The direct comparison to traditional petrochemical plastics further elucidated that bioplastics do not perform better in aspects of chemical complexity and safety (Zimmermann et al. 2020). This is not surprising since, just like conventional plastics, bioplastics rely on the addition of (man-made) chemicals to achieve desired functionalities. In the case of PE and PET, a change in the carbon source of the monomer from petroleum- to plant-based feedstock—so, from a fossil, non-renewable carbon source to a renewable carbon source—does not imply that the plastic's chemistry and processing change. Moreover, bioplastics often fall short in physical and chemical properties compared to their well-established petroleum-based counterparts, due to both material-inherent limitations and still being in the early phase of their development. Consequently, they are blended with other (synthetic) polymers and modified with fillers and (man-made) additives to offset these limitations (Khan et al. 2017). For instance, plasticizers are added to deal with PLA's brittleness and poor thermal properties, and chain extension agents improve the material's impact resistance (Ljungberg and Wesslén 2005; Liu et al. 2013). In addition, pesticides and fertilizers may still be present in bio-based products as a leftover from the production of natural resources (Gironi and Piemonte 2011a).

These findings highlight that chemical safety aspects need stronger consideration in the development of plastic alternatives. To get a better understanding of bioplastics' toxicity, future systematic assessments need to cover all types of bioplastics and all relevant potential human and environmental health implications, since the focus so far has mostly been on PLA and its effects upon environmental release.

Other Sustainability Aspects

Apart from health impacts, other sustainability aspects need to be considered when evaluating the performance of bio-based and bio-degradable materials. These include environmental (e.g., greenhouse gas emission), societal (e.g., competition with food production), and economic (e.g., production costs) aspects of bioplastics' production, use, and management (Gerassimidou et al. 2021).

Life cycle assessments (LCAs) analyze the environmental impacts of a product along its whole life cycle and suggest that bio-based plastics can save fossil resources and reduce greenhouse gas emissions. However, due to the land, water, fertilizers, and pesticide use associated with raw material generation for their production, bio-based plastics can ultimately have a higher societal and environmental impact than conventional plastics (Gironi and Piemonte 2011b). Generally, the outcome of LCAs is highly dependent on the individual products that are being compared, their production, and distribution, as well as on the type and weight of applied assessment criteria, limiting the comparability of different LCAs and making them prone to subjective expert bias (Giovannini et al. 2015). In addition, LCAs completely neglect the toxicity of chemical mixtures associated with the use of conventional plastics or bioplastics

Biodegradable plastics are neither a silver-bullet solution—also not for plastic pollution. The rate of polymeric breakdown largely depends on prevailing ambient conditions, such as temperature, moisture, and the presence of microorganisms that can degrade the plastics. The specific conditions lead to limited biodegradation in most natural environments including home composts and even more so, to a disruption of industrial composting facilities when consumers confuse biodegradable with non-biodegradable plastics (Haider et al. 2019).

And what about oxo-(bio-)degradable plastics? Oxo-(bio-)degradable plastics are a sub-group of biodegradable plastics to which special chemicals, so-called prodegradants or oxo-additives, are added. These additives accelerate natural polymer oxidation and thus, promote degradation. But even oxo-degradable plastics may not fully degrade under conditions prevailing in natural environments, nor under the specified industrial composting conditions (Geueke 2014). Indeed, an early fragmentation of these plastics merely accelerates the formation of microplastics but does not solve the plastic pollution problem. The controversy around their environmental impacts has led to the introduction of bans on oxo-degradable plastics in some countries, e.g., under the European Single-Use Plastics Directive (EU 2019), but in other countries they are still being used.

In general, insufficient or unclear labeling and "green" marketing of biodegradable plastics leave consumers confused about waste management, favoring improper disposal or littering (European Environment Agency 2020). Additionally, biodegradable plastics are mostly designed for single-use and when they become waste, they are recovered energetically (i.e., they are incinerated), conflicting with ambitions towards a circular economy, where resource loss is minimized (more on the circular economy in the dedicated section below). New polymers in general, biodegradable and also bio-based, "contaminate" established recycling schemes since they are not yet adapted to them (Federal Environment Agency 2012).

Consequently, for the time being, biodegradable plastics' suitability is limited to niche applications where (1) biodegradation is part of the function (e.g., in medical applications), (2) environmental entry cannot be avoided and collection is unlikely (e.g., firework cartridges), or where (3) recycling is not practical (e.g., agricultural mulch films). However, more attention has to be paid to chemical safety aspects, especially due to their reintroduction into natural environments. For instance, material design should guarantee that byproducts of degradation are nontoxic and that none of the chemicals which leach during use or after disposal are hazardous or untested. Besides, further ecotoxicity and human health measures need integration in LCAs or composting standards certificates, such as the European standard EN 14432 currently only considering plant germination and growth (DIN 2000).

In contrast to biodegradables, bio-based plastics offer a more promising alternative to conventional plastics with broader applicability. Their ecological footprint can be reduced by using agro-industrial, forest, and food wastes as raw materials originally gained from sustainable agriculture (Álvarez-Chávez et al. 2012). Approaches to enhance the physico-chemical and mechanical properties have been proposed such as the use of algae-based biomass and the inclusion of nanocomposites in polymerization. Besides an improvement of their material properties, technologies for the production and management of bio-based plastics need improvement, for instance, to increase process yields and the energy sufficiency of bioconversion, and to adapt existing recycling infrastructure to these new materials (Ferreira-Filipe et al. 2021).

In conclusion, neither adequate chemical safety[4] nor sustainability will be achieved by just replacing petroleum-based plastics with bio-based and biodegradable materials. Instead, adopting a holistic view, considering environmental, political, social, technical, economic, and also health-related aspects, as well as their dynamic interactions, are essential to avoid regrettable substitutions and develop comprehensive and robust solutions for the future (Gerassimidou et al. 2021).

---

[4] Safety in the sense that plastics do not contain hazardous and/or untested chemicals.

Safe and Circular—Circular Economy

As outlined, just replacing one material with another is not the solution to the plastic pollution problem in general nor the chemical safety issue in particular. A more comprehensive approach is to reduce material and resource use (incl. energy from fossil carbon combustion). The circular economy pursues this target. In an ideal circular economy, materials circulate endlessly to minimize raw material use and, also to reduce waste, energy use and loss, and greenhouse gas emissions along the whole product life-cycle (EPEA 2019). By these means, economic growth is ultimately decoupled from resource consumption which is considered a prerequisite for the survival of humanity on this planet and natural resources can regenerate.

Pillars of a Circular Economy

Several measures support a circular economy. The most favorable one is to avoid the use of industrial products in general or hazardous chemicals in particular wherever possible. One approach here is to ask whether the specific function that a chemical, such as a plastic polymer, provides is essential for the health, safety, and/or functioning of our society. For example, fluoropolymers are used in non-stick cookware but have been considered as a substance of concern due to the release of per- and polyfluoroalkyl substances (PFAS). Thus, one could ask: Is the prevention of food sticking to cookware during food preparation essential for the functioning of society? (Cousins et al. 2021). A second example is the use of microplastics in personal care and cosmetic products as emulsifying or abrasive agents. First, cosmetics as such are not essential for the health or the functioning of our society in general. In addition, microbeads can technically be replaced by safer alternatives as illustrated by the fact that some cosmetic companies have voluntarily renounced the use of microbeads and that many countries have completely banned the use of microbeads in cosmetics. Another approach to reducing the number of man-made products in use is collaborative consumption. Here one accesses a service (e.g., mobility) instead of possessing the object itself (e.g., owning a car). Other pillars of the circular economy include reuse and repair, and re-manufacturing which help to increase the service life of a product. However, service life is not endless, so even reused and repaired products are discarded and replaced at some point.

Recycling

In contrast to reuse, recycled materials re-enter a new life-cycle after one "service-life-cycle." Recycling is often understood as the conversion of one product into a new product serving the same purpose and having the same properties again and again, e.g., one used yoghurt cup is turned into a new yoghurt cup of the same quality. However, for most FCAs or FCMs this so-called "closed-loop" recycling is not possible. Instead, their recycling is "open loop". This means that a material may be re-used to produce another product, which is usually of lower grade and purity. An example is the conversion of a food packaging into a flower pot. In addition, only a small portion of the original resource re-enters the loop while the majority becomes waste. In that way, "open loop" disposal is only postponed but not avoided, and even recycled plastics eventually end up either in landfills or being incinerated (Fedkin n.d.). Plastic recycling is not necessarily "closed-loop" as is reflected by the legal definition for recycling in the EU (EU 2008). Here recycling is defined as "any recovery operation by which waste materials are reprocessed into products, materials, or substances whether for the original or other purposes. It includes the reprocessing of organic material but does not include energy recovery and the reprocessing into materials that are to be used as fuels or for backfilling operations".

Not only is the recycling of plastics such as PET limited to a few (three to five) cycles, but to maintain the material's properties, especially its strength, new materials (also referred to as "virgin materials") have to be blended with the PET recyclate (Schyns and Shaver 2021). While recycled PET due to its relatively high inertness is presently not considered a major safety concern (EFSA 2018), other recycled plastics like the polyolefins (e.g., polyethylene and polypropylene), are. This is because processing does not necessarily destroy or remove contaminants. For example, polyolefins are of lower heat-resistance than PET and less inert meaning that (1) chemicals will migrate at higher levels, that (2) they can absorb more chemicals, including environmental contaminants, and (3) they cannot be heat treated at the same elevated temperatures as PET, to remove volatile contaminants. Thus, oligomers, additives, and their degradation products, as well as flavor compounds and other types of contaminants may remain in the recycled material while further chemicals may be added to compensate for functionality losses (Schyns and Shaver 2021). In addition, during the recycling process, further contaminants from non-food-grade plastics may be introduced into the material (cross-contamination from non-food grade packaging waste disposal) (Dreolin et al. 2019). For instance, waste electric and electronic equipment (WEEE) may illegally be used to cover the demand for black plastics FCAs. Inefficient sorting of WEEE black plastic waste can lead to the contamination of black consumer goods with hazardous substances including brominated flame retardants and heavy metals (Turner 2018). Accordingly, recycled plastics, especially black ones, contain higher levels and numbers of chemicals that can migrate into food. In Europe, a specific regulation (EC No 282/2008) applies to recycled plastic FCMs and FCAs, implying that each recycling process for food-grade plastics must undergo authorization (European Union 2008). The European Food Safety Authority (EFSA) has reviewed plastic recycling processes for FCMs and has considered some as acceptable for food contact. None of these recycling processes for plastic food contact uses has yet been authorized by the European Commission.

For some FCMs, like those made of PET, recycling processes are well established and meet EFSA's safety requirements. For other plastics, recycling processes do not meet safety requirements for use in food contact. Besides the availability of established processes, the recyclability of packaging also strongly depends on the material type and its properties. For example, plastic multilayers consist of several plastic layers and polymers. The combination of the different polymers provides excellent technical properties allowing the packaging and preservation even of quickly perishable products such as fresh meat. However, plastic multilayers cannot be recycled into new food packaging but only non-food grade products. This is because the different layers may not be separable and compatibilizers may be used in their blending (Geueke et al. 2018). Currently, the majority of plastic food packaging is mostly only used once (Geyer et al. 2017).

In contrast to impermanent materials like plastics, permanent materials like glass and metal can be reused and recycled repeatedly without a change of their inherent properties and the addition of primary materials or further chemicals in recycling (Metal Packaging Europe 2018). Thus, the chemical safety of permanent materials remains largely unaffected by recycling. Due to the very high heat used in the remelting of these materials, organic compounds and microorganisms are destroyed. However, also permanent materials are not by definition free of hazardous substances (Geueke et al. 2018). As an example, sands used for manufacturing glass can contain lead still present in the final glass product (Shotyk and Krachler 2007). Therefore, specific methods applied in glass recycling such as UV fluorescence analysis or treatment with acids help to sort out contaminated cullets or to remove heavy metals from the material, respectively. Generally, based on their high inertness, permanent materials have a low propensity for absorption of chemicals (Geueke et al. 2018). This also means that inert materials do not take on smells and dyes (e.g., from their content) which makes reuse possible.

As outlined, especially for non-inert materials such as plastics, one important hindrance of recycling is the presence of chemicals. Both benign substances with a strong smell, as well as hazardous chemicals, may be preventing re-use and recycling.

Challenges Towards Non-Toxic Plastics Recycling

To achieve non-toxic plastics recycling, several challenges have to be overcome. One challenge is the currently limited information on the chemicals (of concern) added in previous life cycles and available to waste handling and recycling institutions. Another challenge is that once legal substances, now forbidden due to demonstrated risk, may still be contained in recycled plastics ("legacy contaminants", see also Eckert in this volume). Furthermore, when a product becomes waste, the waste legislation applies in the EU and "end-of-waste criteria" have to be met. However, waste streams and recovery processes are complex and waste management rules do not apply equally across the EU's Member States, meaning regulations are not harmonized. Thus, legal circumstances are unclear on how waste becomes a new material or product in the EU. In addition, the lacking alignment of EU-wide rules on waste and regulations on products can impact the safety of the secondary raw material. As an example, flexible PVC waste is classified as non-hazardous while the product recovered from it is classified as a hazardous chemical mixture (EC 2018).

Consequently, there is a need to harmonize and adapt existing EU regulations on chemicals (e.g., waste directive, REACH, regulation on FCMs) to support non-toxic cycles of plastics and other materials. New regulatory frameworks should further account for the overall migration and potential toxicity of the chemical mixtures present in the final (recycled) product. A reduction in complexity, using fewer polymers and additives, may favor the chemical safety of the primary and the recycled product while simultaneously increasing recycling rates as a whole.

Not only chemical safety aspects limit reuse and recycling but also technical aspects. Efficient collection, separation, and cleaning processes build the prerequisite for recycling and reuse, as does the demand on the market. However, respective infrastructure is missing in many countries and the demand favors virgin over reused and recycled plastics, as these are currently both cheaper and easier for establishing regulatory compliance.

Is a Circular Economy THE Solution?

Even when reuse and recycling schemes improve and transition to a circular economy progresses, it will not solve all issues connected with plastic packaging use. The current way plastics are consumed will always be associated with externalities affecting human and/or environmental health. One way in which plastics have been described is as a planetary boundary threat. Planetary boundaries define the safe space for humanity to operate in without causing unacceptable environmental changes that in turn lead to a destabilization of the Earth's ecosystems and threaten humanity's livelihood on the planet (Rockström et al. 2009). Plastics transgress these thresholds because they are persistent pollutants that cause irreversible exposure on the global scale and have a currently not fully known disruptive effect on a vital process of the Earth system. Based on this fact and the pertinent knowledge gaps, it has been proposed that plastics should be candidates for a precautionary phaseout (Jahnke et al. 2017). This illustrates the far-reaching implications associated with plastic use and leads to the question of whether the elimination of plastics would be THE solution. In some cases, a plastic product is not needed, for instance, in the packaging of sliced fruits when whole fruit could be sold without any packaging instead. In other cases, an alternative material to plastics having no or a lower impact may be identified. However, some uses of plastics are of very high societal importance and can be considered essential uses. Therefore, a complete phaseout of all plastics seems unfeasible from a current perspective, although this should still be the overall goal when stakeholders are thinking about and developing solutions. Due to the complexity of all problems associated with plastic use, there is no one-size-fits-all solution but instead, multiple strategies, adapted to specific circumstances, a certain product, and its application, need developing and implementing. Importantly, reducing society's overall consumption of plastics implies a change of current behavior and business models, e.g., towards producing and buying locally. These are important aspects to consider as regulators move to taxing plastics and subsidizing recycling schemes—especially the latter may lead to the unintended consequence of a technology lock-in, where the actual goal of reducing plastics production and consumption is missed.

Conclusion

Plastics, especially used for food packaging, are highly useful and economically advantageous materials. They can contribute to extending the shelf-life of food products and thereby enable highly profitable, globalized business models in food manufacture and retail. Plastics for food packaging also enable a lifestyle of amazing convenience.

But plastics are chemically highly complex, and some of the chemical constituents in plastics transfer into foods. This means that almost the entire human population is exposed to plastic chemicals on a daily basis. Many of these chemicals have never been assessed for their risk of harming human health or the environment. Nor are the mixtures of chemicals that migrate from plastics into foods well characterized for their chemical composition and toxicological consequences. Indeed, all plastics contain NIAS, which most often remain unknown.

The way that chemical safety is assessed for plastics assumes that for all synthetic, man-made chemicals there is an exposure level below which no harm to human health is caused ("the dose makes the poison"). But this assumption, while certainly valid for some chemicals, has been disproved by modern science for many cases such as carcinogens and endocrine-disrupting chemicals. Also, it is not valid when humans ingest mixtures of synthetic chemicals that leach from plastics into foodstuffs. Consequently, chemical safety assessments are not based on today's knowledge of how chemicals interact with biological targets and contribute to chronic diseases in the entire human population. Therefore, updates in regulatory approaches for assessing the chemical safety of plastics are urgently needed. Reducing exposure of the general population even to low levels of known hazardous chemicals from plastics will lead to large benefits to society as a whole.

Another key issue is the need to address plastic chemicals and toxicity when discussing plastic packaging waste management options. Chemicals can accumulate in plastics during recycling, and some types of plastics are more suitable for recycling in food contact applications than others. However, because plastics are a planetary boundary threat due to their environmental persistence, there is a need for reducing the overall production and consumption of plastics. Therefore, investments into plastics recycling infrastructure must be made with caution to avoid long-term technological lock-in situations.

With the increased awareness of the problems associated with plastic consumption, work on solutions has also increased. However, developing and implementing solutions to complex problems, such as plastic pollution and plastic-associated health impacts, is not an easy task, and it has to be avoided that "today's solutions become tomorrow's problems." For example, looking back, the replacement of glass with plastics has facilitated distribution (since unbreakable) and reduced cost (cheap production) but also made high production and consumption of packaging possible and has eventually provided us with today's plastic-associated problems. Therefore, it is key to take a holistic approach to problem-solving. This means to not simply focus on one symptom of the problem, such as plastic waste/pollution, but consider all impacts associated with a plastic product's life cycle. For instance, bio-based plastics may be beneficial in terms of saving fossil resources but may have negative consequences on societies (e.g., land consumption) or biodiversity (e.g., pesticides), when not considered holistically. Only when involving all stakeholders in problem-solving and when all acknowledge their responsibility, can regrettable, small-scale and end-of-pipe solutions be avoided. There will be no simple one-size-fits-all solution; instead, multiple strategies will be required, that are customized to a product and the local circumstances.

References

Adhikari, Dinesh, Masaki Mukai, Kenzo Kubota, Takamitsu Kai, Nobuyuki Kaneko, Kiwako S. Araki, and Motoki Kubo (2016). Degradation of bioplastics in soil and their degradation effects on environmental microorganisms. *Journal of Agricultural Chemistry and Environment*, 5, 23–34.

Álvarez-Chávez, Clara R., Sally Edwards, Rafael Moure-Eraso, and Kenneth Geiser (2012). Sustainability of bio-based plastics: General comparative analysis and recommendations for improvement. *Journal of Cleaner Production*, 23, 47–56.

Asensio, Esther, Laura Montañés, and Cristina Nerín (2020). Migration of volatile compounds from natural biomaterials and their safety evaluation as food contact materials. *Food and Chemical Toxicology*, 142, 111457.

ATSDR (2016). *Addendum to the toxicological profile for vinyl chloride*. Agency for Toxic Substances and Disease Registry Division of Toxicology and Human Health Sciences Atlanta, GA 30329-4027.

Aznar, Margarita, Sara Ubeda, Nicola Dreolin, and Cristina Nerín (2019). Determination of non-volatile components of a biodegradable food packaging material based on polyester and polylactic acid (PLA) and its migration to food simulants. *Journal of Chromatography A*, 1583, 1–8.

Bach, Cristina, Xavier Dauchy, Isabelle Severin, Jean-François Munoz, Serge Etienne, and Marie-Christine Chagnon (2013). Effect of temperature on the release of intentionally and non-intentionally added substances from polyethylene terephthalate (PET) bottles into water: Chemical analysis and potential toxicity. *Food Chemistry*, 139, 672–80.

Balestri, Elena, Virginia Menicagli, Viviana Ligorini, Sara Fulignati, Anna M. Raspolli Galletti, and Claudio Lardicci (2019). Phytotoxicity assessment of conventional and biodegradable plastic bags using seed germination test. *Ecological Indicators*, 102, 569–80.

Breton, Carrie V., Remy Landon, Linda G. Kahn, Michelle B. Enlow, Alicia K. Peterson, Theresa Bastain, Joseph Braun, Sarah S. Comstock, Cristiane S. Duarte, Alison Hipwell, Hong Ji, Janine M. LaSalle, Rachel L. Miller, Rashelle Musci, Jonathan Posner, Rebecca Schmidt, Shakira F. Suglia, Irene Tung, Daniel Weisenberger, Yeyi Zhu, and Rebecca Fry (2021). Exploring the evidence for epigenetic regulation of environmental influences on child health across generations. *Communications Biology*, 4, 769.

Bschir, Karim (2017). Risk, uncertainty and precaution in science: The threshold of the toxicological concern approach in food toxicology. *Science and Engineering Ethics*, 23, 489–508.

Calafat, Antonia M., Xiaoyun Ye, Lee-Yang Wong, John A. Reidy, and Larry L. Needham (2008). Exposure of the U.S. population to bisphenol A and 4-tertiary-octylphenol: 2003–2004. *Environmental Health Perspectives*, 116, 39–44.

Castle, Laurence, David Price, and John V. Dawkins (1996). Oligomers in plastics packaging. Part 1: Migration tests for vinyl chloride tetramer. *Food Additives and Contaminants*, 13, 307–14.

Costa, Sofia A., Sofia Vilela, Daniela Correia, Milton Severo, Carla Lopes, and Duarte Torres (2020). Consumption of packaged foods by the Portuguese population: Type of materials and its associated factors. *British Food Journal*, 123, 833–46.

Cousins, Ian T., Jamie C. de Witt, Juliane Glüge, Gretta Goldenman, Dorte Herzke, Rainer Lohmann, Mark Miller, Carla A. Ng, Sharyle Patton, Martin Scheringer, Xenia Trier, and Zhanyun Wang (2021). Finding essentiality feasible: Common questions and misinterpretations concerning the "essential-use" concept. *Environmental Science. Processes & Impacts*, 23 (8), 1079–1087.

Deichmann, W. B., D. Henschler, B. Holmstedt, and G. Keil (1986). What is there that is not poison? A study of the Third Defense by Paracelsus. *Archives of Toxicology*, 58, 207–213.

DIN (2000). Packaging - Requirements for packaging recoverable through composting and biodegradation - Test scheme and evaluation criteria for the final acceptance of packaging; EN 13432:2000. 06.11.2020 https://www.din.de/en/getting-involved/standards-committees/navp/publications/wdc-beuth:din21:32115376.

Dolinoy, Dana C., Dale Huang, and Randy L. Jirtle (2007). Maternal nutrient supplementation counteracts bisphenol A-induced DNA hypomethylation in early development. *Proceedings of the National Academy of Sciences of the United States of America*, 104, 13056–13061.

Dreolin, Nicola, Margarita Aznar, Sabrina Moret, and Cristina Nerin (2019). Development and validation of a LC-MS/MS method for the analysis of bisphenol a in polyethylene terephthalate. *Food Chemistry*, 274, 246–253.

Ebner, Ingo, Steffi Haberer, Stefan Sander, Oliver Kappenstein, Andreas Luch, and Torsten Bruhn (2020). Release of melamine and formaldehyde from melamine-formaldehyde plastic kitchenware. *Molecules*, 25 (16), 3629.

EC (2018). Communication from the commission to the European Parliament, the Council, the European Economic and Social Committee and the Commitee of the Regions on the implementation of the circular economy package: *Options to address the interface between chemical, product and waste legislation*. Strasbourg. 02.09.2021 https://eur-lex.europa.eu/legal-content/EN/TXT/HTML/?uri=CELEX:52018DC0032&from=EN.

Eckert, Sandra (2022). Circularity in the plastic age: Policymaking and industry action in the European Union. In Johanna Kramm and Carolin Völker (eds.). *Living in the plastic age. Perspectives from humanities, social sciences and environmental sciences*. Frankfurt, New York: Campus.

EFSA (2018). Register of questions. 06.01.2021 http://registerofquestions.efsa.europa.eu/.

EPEA (2019). Von der Wiege zur Wiege. Produktionsprozesse neu denken. Hamburg. 06.01.2021 https://epea.com/ueber-uns/cradle-to-cradle.

Ernstoff, Alexi, Monia Niero, Jane Muncke, Xenia Trier, Ralph K. Rosenbaum, Michael Hauschild, and Peter Fantke (2019). Challenges of including human exposure to chemicals in food packaging as a new exposure pathway in life cycle impact assessment. *The International Journal of Life Cycle Assessment*, 24, 543–52.

EU (2004). Regulation (EC) No 1935/2004 of the European Parliament and the Council of 27 October 2004 on materials and articles intended to come into contact with food. 31.10.2018.

EU (2008). Directive 2008/98/EC of the European Parliament and of the Council of 19 November 2008 on waste and repealing certain Directives. 14.11.2020.

EU (2019). Directive (EU) 2019/904 of the European Parliament and of the Council of 5 June 2019 on the reduction of the impact of certain plastic products on the environment. 04.02.2020.

European Bioplastics (2020). Bioplastics market data 2020. 19.08.2021 https://www.european-bioplastics.org/market/.

European Environment Agency (2020). *Biodegradable and compostable plastics— Challenges and opportunities.* Copenhagen, Denmark. 23.08.2021.

European Parliament (2016). European Parliament resolution of 6 October 2016 on the implementation of the Food Contact Materials Regulation (EC) No 1935/2004 (2015/2259(INI)). P8_TA-PROV(2016)0384 (pdf). 27.09.2021.

European Union (2008). Commission regulation (EC) No 282/2008 of 27 March 2008 on recycled plastic materials and articles intended to come into contact with foods and amending Regulation (EC) No 2023/2006. 03.11.2020.

Federal Environment Agency (2012). Study of the environmental impacts of packagings made of biodegradable plastics. Dessau-Roßlitz. 06.01.2021 https://www.umweltbundesamt.de/publikationen/study-of-environmental-impacts-of-packagings-made.

Fedkin, M. V. (n.d.). EME 807: Technologies for sustainability systems. 14.09.2021 https://www.e-education.psu.edu/eme807/node/624

Feil, Robert, and Mario F. Fraga (2012). Epigenetics and the environment: Emerging patterns and implications. *Nature Reviews. Genetics*, 13, 97–109.

Ferreira-Filipe, Diogo A., Ana Paço, Armando C. Duarte, Teresa Rocha-Santos, and Ana L. Patrício Silva (2021). Are biobased plastics green alternatives?-A critical review. *International Journal of Environmental Research and Public Health*, 18 (15), 7729.

Food Additive Amendment (1958). Food for human consumption. U.S. Code of Federal Regulations. 21 CFR 170.3. 06.09.2021.

Gerassimidou, Spyridoula, Olwenn V. Martin, Stephen P. Chapman, John N. Hahladakis, and Eleni Iacovidou (2021). Development of an integrated sustainability matrix to depict challenges and trade-offs of introducing bio-based plastics in the food packaging value chain. *Journal of Cleaner Production*, 286, 125378.

Geueke, B. (2014). Dossier – Bioplastics as food contact materials. Food Packaging Forum. https://doi.org/10.5281/zenodo.33517

Geueke, Birgit, Ksenia Groh, and Jane Muncke (2018). Food packaging in the circular economy: Overview of chemical safety aspects for commonly used materials. *Journal of Cleaner Production*, 193, 491–505.

Geyer, Roland, Jenna R. Jambeck, and Kara L. Law (2017). Production, use, and fate of all plastics ever made. *Science Advances*, 3, e1700782.

Giovannini, Antonio, Michele Dassisti, Francesca Intini, Osiris Canciglieri, and Eduardo Rocha Loures (2015). Life-cycle assessment comparability: A perspective to analyse the heterogeneity. Paisley, SEEP Conference 2015. 10.02.2021.

Gironi, Fausto, and Vincenzo Piemonte (2011a). Bioplastics and petroleum-based plastics: Strengths and weaknesses. *Energy Sources, Part A: Recovery, Utilization, and Environmental Effects*, 33, 1949–1959.

Gironi, Fausto, and Vincenzo Piemonte (2011b). Life cycle assessment of polylactic acid and polyethylene terephthalate bottles for drinking water. *Environmental Progress & Sustainable Energy*, 30, 459–468.

Gore, A. C., V. A. Chappell, S. E. Fenton, J. A. Flaws, A. Nadal, G. S. Prins, J. Toppari, and R. T. Zoeller (2015). Executive summary to EDC-2: The Endocrine Society's second scientific statement on endocrine-disrupting chemicals. *Endocrine Reviews*, 36, 593–602.

Göttermann, S., C. Bonten, A. Kloeppel, S. Kaiser, and F. Brümme (2015). Marine littering-Effects and degradation. *Conference Paper 24. Stuttgarter Kunststoffkolloquium*.

Groh, Ksenia J., Thomas Backhaus, Bethanie CarneyAlmroth, Birgit Geueke, Pedro A. Inostroza, Anna Lennquist, Heather A. Leslie, Maricel Maffini, Daniel Slunge, Leonardo Trasande, A. M. Warhurst, and Jane Muncke (2019). Overview of known plastic packaging-associated chemicals and their hazards. *The Science of the Total Environment*, 651, 3253–3268.

Groh, Ksenia J., Birgit Geueke, Olwenn Martin, Maricel Maffini, and Jane Muncke (2020). Overview of intentionally used food contact chemicals and their hazards. *Environment International*, 150, 106225.

Groh, Ksenia J., and Jane Muncke (2017). *In vitro* toxicity testing of food contact materials: State-of-the-art and future challenges. *Comprehensive Reviews in Food Science and Food Safety*, 16, 1123–1150.

Haider, Tobias P., Carolin Völker, Johanna Kramm, Katharina Landfester, and Frederik R. Wurm (2019). Plastics of the future? The impact of biodegradable polymers on the environment and on society. *Angewandte Chemie International Edition*, 58, 50–62.

Huerta-Lwanga, Esperanza, Jorge Mendoza-Vega, Oriana Ribeiro, Henny Gertsen, Piet Peters, and Violette Geissen (2021). Is the polylactic acid fiber in green compost a risk for *Lumbricus terrestris* and *Triticum aestivum*? *Polymers*, 13 (5), 703.

IARC (2018). Carcinogenicity of quinoline, styrene, and styrene-7,8-oxide. IARC Monographs on the Evaluation of Carcinogenic Risks to Humans Volume 121. IARC, Lyon.

Jahnke, Annika, Hans P. H. Arp, Beate I. Escher, Berit Gewert, Elena Gorokhova, Dana Kühnel, Martin Ogonowski, Annegret Potthoff, Christoph Rummel, Mechthild Schmitt-Jansen, Erik Toorman, and Matthew MacLeod (2017). Reducing uncertainty and confronting ignorance about the possible impacts of weathering plastic in the marine environment. *Environmental Science & Technology Letters*, 4, 85–90.

Kato, Lilian S., and Carlos A. Conte-Junior (2021). Safety of plastic food packaging: The challenges about non-intentionally added substances (NIAS) discovery, identification and risk assessment. *Polymers*, 13 (13), 2077.

Khan, Bahram, Muhammad Bilal Khan Niazi, Ghufrana Samin, and Zaib Jahan (2017). Thermoplastic starch: A possible biodegradable food packaging material - a review. *Journal of Food Process Engineering*, 40, e12447.

Koster, Sander, Alan R. Boobis, Richard Cubberley, Heli M. Hollnagel, Elke Richling, Tanja Wildemann, Gunna Würtzen, and Corrado L. Galli (2011). Application of the TTC concept to unknown substances found in analysis of foods. *Food and Chemical Toxicology*, 49, 1643–1660.

Koster, Sander, Monique Rennen, Winfried Leeman, Geert Houben, Bas Muilwijk, Frederique van Acker, and Lisette Krul (2014). A novel safety assessment strategy for non-intentionally added substances (NIAS) in carton food contact materials. *Food Additives & Contaminants. Part A, Chemistry, Analysis, Control, Exposure & Risk Assessment*, 31, 422–443.

Lambert, Scott, and Martin Wagner (2017). Environmental performance of bio-based and biodegradable plastics: The road ahead. *Chemical Society Reviews*, 46, 6855–6871.

Lithner, Delilah, Jeanette Damberg, Goran Dave, and Ke Larsson (2009). Leachates from plastic consumer products - screening for toxicity with *Daphnia magna*. *Chemosphere*, 74, 1195–2000.

Lithner, Delilah, Ake Larsson, and Goran Dave (2011). Environmental and health hazard ranking and assessment of plastic polymers based on chemical composition. *The Science of the Total Environment*, 409, 3309–3324.

Liu, Wei, Hangquan Li, Xiangdong Wang, Zhongjie Du, and Chen Zhang (2013). Effect of chain extension on the rheological property and thermal behaviour of poly(lactic acid) foams. *Cellular Polymers*, 32, 343–368.

Ljungberg, Nadia, and Bengt Wesslén (2005). Preparation and properties of plasticized poly(lactic acid) films. *Biomacromolecules*, 6, 1789–1796.

Maffini, Maricel V., Birgit Geueke, Ksenia Groh, Bethanie Carney Almroth, and Jane Muncke (2021). Role of epidemiology in risk assessment: A case study of five ortho-phthalates. *Environmental Health*, 20, 114.

Maisanaba, Sara, Silvia Pichardo, María Jordá-Beneyto, Susana Aucejo, Ana M. Cameán, and Ángeles Jos (2014). Cytotoxicity and mutagenicity studies on migration extracts from nanocomposites with potential use in food packaging. *Food and Chemical Toxicology*, 66, 366–372.

McCombie, Gregor (2018). *Enforcement's Perspective*. 27.09.2021 https://ec.europa.eu/food/system/files/2018-09/cs_fcm_eval-workshop_20180924_pres07.pdf.

McCombie, Gregor, Karsten Hötzer, Jürg Daniel, Maurus Biedermann, Angela Eicher, and Koni Grob (2016). Compliance work for polyolefins in food contact: Results of an official control campaign. *Food Control*, 59, 793–800.

Menicagli, Virginia, Elena Balestri, and Claudio Lardicci (2019). Exposure of coastal dune vegetation to plastic bag leachates: A neglected impact of plastic litter. *The Science of the Total Environment*, 683, 737–748.

Mertl, Johannes, Christian Kirchnawy, Veronica Osorio, Angelika Grininger, Alexander Richter, Johannes Bergmair, Michael Pyerin, Michael Washüttl, and Manfred Tacker (2014). Characterization of estrogen and androgen activity of food contact materials by different *in vitro* bioassays (YES, YAS, ERα and AR CALUX) and chromatographic analysis (GC-MS, HPLC-MS). *PLoS One*, 9, e100952.

Metal Packaging Europe (2018). Permanent Materials. 02.09.2021 https://www.metalpackagingeurope.org/sites/default/files/2018-04/Permanent%20Material%20-%20Report%20Carbotech.pdf.

Montévil, Maël, Nicole Acevedo, Cheryl M. Schaeberle, Manushree Bharadwaj, Suzanne E. Fenton, and Ana M. Soto (2020). A combined morphometric and statistical approach to assess nonmonotonicity in the developing mammary gland of rats in the CLARITY-BPA Study. *Environmental Health Perspectives*, 128, 57001.

Muncke, Jane (2021). Tackling the toxics in plastics packaging. *PLoS Biology*, 19, e3000961.

Muncke, Jane, Anna-Maria Andersson, Thomas Backhaus, Justin M. Boucher, Bethanie Carney Almroth, Arturo Castillo Castillo, Jonathan Chevrier, Barbara A. Demeneix, Jorge A. Emmanuel, Jean-Baptiste Fini, David Gee, Birgit Geueke, Ksenia Groh, Jerrold J. Heindel, Jane Houlihan, Christopher D. Kassotis, Carol F. Kwiatkowski, Lisa Y. Lefferts, Maricel V. Maffini, Olwenn V. Martin, John P. Myers, Angel Nadal, Cristina Nerin, Katherine E. Pelch, Seth R. Fernández, Robert M. Sargis, Ana M. Soto, Leonardo Trasande, Laura N. Vandenberg, Martin Wagner, Changqing Wu, R. T. Zoeller, and Martin Scheringer (2020). Impacts of food contact chemicals on human health: A consensus statement. *Environmental Health*, 19, 25.

Muncke, Jane, Thomas Backhaus, Birgit Geueke, Maricel V. Maffini, Olwenn V. Martin, John P. Myers, Ana M. Soto, Leonardo Trasande, Xenia Trier, and Martin Scheringer (2017). Scientific challenges in the risk assessment of food contact materials. *Environmental Health Perspectives*, 125, 95001.

Pieke, Eelco N., Jørn Smedsgaard, and Kit Granby (2018). Exploring the chemistry of complex samples by tentative identification and semiquantification: A food contact material case. *Journal of Mass Spectrometry JMS*, 53, 323–335.

Pinter, Elisabeth, Bernhard Rainer, Thomas Czerny, Elisabeth Riegel, Benoît Schilter, Maricel Marin-Kuan, and Manfred Tacker (2020). Evaluation of the suitability of mammalian *in vitro* assays to assess the genotoxic potential of food contact materials. *Foods*, 9 (2), 237.

PlasticsEurope (2019). Plastics – the facts 2019: An analysis of European plastics production, demand and waste data. 09.11.2020 https://www.plasticseurope.org/de/resources/publications/2154-plastics-facts-2019.

Poças, M.F.F., J. C. Oliveira, H. J. Pinto, M. E. Zacarias, and T. Hogg (2009). Characterization of patterns of food packaging usage in Portuguese homes. *Food Additives & Contaminants: Part A*, 26, 1314–1324.

Qian, Shasha, Hanxu Ji, XiaoXiao Wu, Ning Li, Yang Yang, Jiangtao Bu, Xiaoming Zhang, Ling Qiao, Henglin Yu, Ning Xu, and Chi Zhang (2018). Detection and quantification analysis of chemical migrants in plastic food contact products. *PLoS One*, 13, e0208467.

Robles-Matos, Nicole, Tre Artis, Rebecca A. Simmons, and Marisa S. Bartolomei (2021). Environmental exposure to endocrine disrupting chemicals influences genomic imprinting, growth, and metabolism. *Genes*, 12 (8), 1153.

Rockström, Johan, Will Steffen, Kevin Noone, Asa Persson, F. S. Chapin, Eric F. Lambin, Timothy M. Lenton, Marten Scheffer, Carl Folke, Hans J. Schellnhuber, Björn Nykvist, Cynthia A. de Wit, Terry Hughes, Sander van der Leeuw, Henning Rodhe, Sverker Sörlin, Peter K. Snyder, Robert Costanza, Uno Svedin, Malin Falkenmark, Louise Karlberg, Robert W. Corell, Victoria J. Fabry, James Hansen, Brian Walker, Diana Liverman, Katherine Richardson, Paul Crutzen, and Jonathan A. Foley (2009). A safe operating space for humanity. *Nature*, 461, 472–475.

Rose, Geoffrey (1981). Strategy of prevention: Lessons from cardiovascular disease. *British Medical Journal (Clinical Research Ed.)*, 282, 1847–1851.

Schyns, Zoé O. G., and Michael P. Shaver (2021). Mechanical recycling of packaging plastics: A review. *Macromolecular Rapid Communications*, 42, e2000415.

Severin, Isabelle, Emilie Souton, Laurence Dahbi, and Marie C. Chagnon (2017). Use of bioassays to assess hazard of food contact material extracts: State of the art. *Food and Chemical Toxicology*, 105, 429–447.

Shotyk, William, and Michael Krachler (2007). Lead in bottled waters: Contamination from glass and comparison with pristine groundwater. *Environmental Science & Technology*, 41, 3508–3513.

Silva, Manori J., Dana B. Barr, John A. Reidy, Nicole A. Malek, Carolyn C. Hodge, Samuel P. Caudill, John W. Brock, Larry L. Needham, and Antonia M. Calafat (2004). Urinary levels of seven phthalate metabolites in the U.S. population from the National Health and Nutrition Examination Survey (NHANES) 1999-2000. *Environmental Health Perspectives*, 112, 331–338.

Simoneau, Catherine (2015). Annual report 2014 of the EURL-FCM on activities carried out for the implementation of Regulation (EC) no 882/2004. Luxembourg. 22.07.2022 https://publications.jrc.ec.europa.eu/repository/handle/JRC95289

Souza, Patrícia M. S., Nádia A. Corroqué, Ana R. Morales, Maria A. Marin-Morales, and Lucia H. I. Mei (2013). PLA and organoclays nanocomposites: Degradation process and evaluation of ecotoxicity using *Allium cepa* as test organism. *Journal of Polymers and the Environment*, 21, 1052–1063.

Takano, Hirohisa, Rie Yanagisawa, Ken-ichiro Inoue, Takamichi Ichinose, Kaori Sadakane, and Toshikazu Yoshikawa (2006). Di-(2-ethylhexyl) phthalate enhances atopic dermatitis-like skin lesions in mice. *Environmental Health Perspectives*, 114, 1266–1269.

Tanner, Eva M., Maria U. Hallerbäck, Sverre Wikström, Christian Lindh, Hannu Kiviranta, Chris Gennings, and Carl-Gustaf Bornehag (2020). Early prenatal exposure to suspected endocrine disruptor mixtures is associated with lower IQ at age seven. *Environment International*, 134, 105185.

Turner, Andrew (2018). Black plastics: Linear and circular economies, hazardous additives and marine pollution. *Environment International*, 117, 308–318.

van Poucke, Christof, Christ'l Detavernier, Jan F. van Bocxlaer, Rudi Vermeylen, and Carlos van Peteghem (2008). Monitoring the benzene contents in soft drinks using headspace gas chromatography-mass spectrometry: A survey of the situation on the belgian market. *Journal of Agricultural and Food Chemistry*, 56, 4504–4510.

Vandenberg, Laura N., Theo Colborn, Tyrone B. Hayes, Jerrold J. Heindel, David R. Jacobs, Duk-Hee Lee, Toshi Shioda, Ana M. Soto, Frederick S. vom Saal, Wade V. Welshons, R. T. Zoeller, and John P. Myers (2012). Hormones and endocrine-disrupting chemicals: Low-dose effects and nonmonotonic dose responses. *Endocrine Reviews*, 33, 378–455.

Vandenberg, Laura N., Patricia A. Hunt, and Andrea C. Gore (2019). Endocrine disruptors and the future of toxicology testing - lessons from CLARITY-BPA. *Nature reviews. Endocrinology*, 15, 366–374.

Wagner, Martin, and Jörg Oehlmann (2009). Endocrine disruptors in bottled mineral water: Total estrogenic burden and migration from plastic bottles. *Environmental Science and Pollution Research International*, 16, 278–286.

Wagner, Martin, Michael P. Schlüsener, Thomas A. Ternes, and Jörg Oehlmann (2013). Identification of putative steroid receptor antagonists in bottled water: Combining bioassays and high-resolution mass spectrometry. *PLoS One*, 8, e72472.

Wang, Aolin, Dimitri P. Abrahamsson, Ting Jiang, Miaomiao Wang, Rachel Morello-Frosch, June-Soo Park, Marina Sirota, and Tracey J. Woodruff (2021). Suspect screening, prioritization, and confirmation of environmental chemicals in maternal-newborn pairs from San Francisco. *Environmental Science & Technology*, 55, 5037–5049.

Wang, Yufei, and Haifeng Qian (2021). Phthalates and their impacts on human health. *Healthcare*, 9 (5), 603.

Yang, Chun Z., Stuart I. Yaniger, V. C. Jordan, Daniel J. Klein, and George D. Bittner (2011). Most plastic products release estrogenic chemicals: A potential health problem that can be solved. *Environmental Health Perspectives*, 119, 989–996.

Zimmermann, Lisa, Zdenka Bartosova, Katharina Braun, Jörg Oehlmann, Carolin Völker, and Martin Wagner (2021). Plastic products leach chemicals that induce *in vitro* toxicity under realistic use conditions. *Environmental Science & Technology*, 55 (17), 11814–11823.

Zimmermann, Lisa, Georg Dierkes, Thomas A. Ternes, Carolin Völker, and Martin Wagner (2019). Benchmarking the in vitro toxicity and chemical composition of plastic consumer products. *Environmental Science & Technology*, 53, 11467–11477.

Zimmermann, Lisa, Andrea Dombrowski, Carolin Völker, and Martin Wagner (2020). Are bioplastics and plant-based materials safer than conventional plastics? *In vitro* toxicity and chemical composition. *Environment International*, 145, 106066.

Zota, Ami R., Antonia M. Calafat, and Tracey J. Woodruff (2014). Temporal trends in phthalate exposures: Findings from the National Health and Nutrition Examination Survey, 2001–2010. *Environmental Health Perspectives*, 122, 235–241.

Circularity in the Plastic Age: Policymaking and Industry Action in the European Union

Sandra Eckert

Introduction

Environmental policy in the member states of the European Union (EU) is to a large extent determined by supranational rather than national policy choices, with up to around 80 percent of national environmental legislation originating in EU law (Töller 2010). The environmental issues posed by plastics throughout their lifecycle do not form an exception to this pattern. Regulatory activity of European policymakers with regards to the environmental impact of plastics has proven particularly intense over the past decade. Several policy measures have been taken in the context of the circular economy (CE) agenda which has the objective of decoupling growth from resource use and was launched in 2014. Sustainability issues around plastics are indeed pressing both at the resource extraction stage as well as at the end of life cycle, while health issues pose concern in the usage phase. At the stage of resource extraction, it is the transition to a low-carbon economy that poses a tremendous challenge to the fossil-based industry. The global production of conventional plastics continues to rise, which also means that the input of fossil fuels sees an upward trend. Production went up from 15 million tons produced in 1964 to 360 million tons produced in 2017 (PlasticsEurope 2019a; World Economic Forum et al. 2016, 25). The forecast for the share of global oil production going into plastics production amounts to 20 percent by 2050 in comparison to around eight percent in 2016 (PlasticsEurope 2019a; World Economic Forum et al. 2016). At the end of life cycle, only a small proportion of plastics goes into recycling. In Europe plastics production amounted to 61.8 million tons in 2018, of which 29.1 million tons were collected, and 9.4 million tons were sent to recycling, that is around 15 percent of total production (PlasticsEurope 2019a). In the use stage controversies center around consumers' exposure to toxic substances, an issue at the heart of the mobilization of environmental groups

(Eckert 2019, 93–100). Public communication by industry, by contrast, focuses on the benefits of plastic products, such as lightweight or thermic qualities, which are frequently attributed to desirable goals such as saving energy or avoiding food waste (Eckert 2021).

In this context, the CE approach both holds potential and presents risks: it has potential as it could allow to effectively tackle the issues with regard to the material's sustainability throughout its lifecycle; while it is risky because it opens up possibilities for industry co-optation and a shift of policy focus to the less problematic life cycle stages. This is even more strongly the case given that the EU's plastics strategy not only forms a key component of the circular economy agenda (European Commission 2018b), but is also supposed to serve as a role model for the regulation of other resource intense sectors (European Commission 2020a). In view of these challenges, this chapter addresses the following questions: To what extent does the CE policy agenda facilitate the transformation towards more sustainability of plastics? How does the CE approach affect the policy-industry nexus?

To address these questions the chapter is structured as follows: First, it presents the CE concept as one possible policy angle on plastics. It then revisits the trajectory of EU policies on plastics in a second part, engaging with the earlier debate on PVC as an illustrative case for the current controversies on plastics in the CE. Finally, it provides a critical appraisal of the merits of circularity to address plastics, as well as the role of industry in the policy process. The discussion will show that a comprehensive material-specific approach has emerged over the past decade while before that time measures such as those on PVC were adopted as a targeted response to policy controversies and public pressure. It will also highlight that the policy-industry nexus has been consolidated with the CE approach.

Revisiting the Policy-Industry Nexus in the Circular Economy

The CE has become a key concept in rethinking industrial practice and systems. It guides the environmental policy agenda in many constituencies across the globe (McDowall et al. 2017). The CE concept seeks to tackle environmental issues of materials and products throughout their lifecycle. As mentioned already, such an approach also bears the risk of industry co-optation where it allows actors to shift attention away from more problematic lifecycle stages. The following sections will revisit the rich and growing CE literature, and then move on to discuss the role of regulation and industry action in the pursuit of the CE.

The "Re-Words" in the Growing CE Literature

The literature on the CE has grown rapidly and has generated divergent definitions (Blomsma and Brennan 2017; Geissdoerfer et al. 2017; Murray et al. 2017; Reike et al. 2018). Focusing on various "re-words" such as "re-cycling", "re-duce" or "re-use" can be a good way to structure the debate and consider its insights for plastics. Reike and colleagues (2018, 253–254) identify as many as 38 "re-words" in the literature, and find that authors tend to conceptualize the CE making reference to three to up to ten "re-words". Moreover, they conclude from their stocktaking that many contributions lack conceptual clarity, or fail to establish a hierarchy for the measures these "re-words" refer to (Reike et al. 2018, 253–254). In view of these considerations, I will rely on the work of Potting and colleagues (2017, 5) who have approached the concept as a continuum of ten "re-words". They rank them with respect to the level of ambition of the measures involved in a space between a linear economy and a truly circular economy (see Fig. 1). This conceptualization will be a helpful point of reference in understanding the changing focus and level of ambition of the European policy agenda.

Level of ambition	CE Continuum	10 "re-words"
CIRCULAR *(Increasing circularity ↑)*	Smarter production, use, and manufacture	refuse
		rethink
		reduce
	Extend lifespan of products and its parts	reuse
		repair
		refurbish
		remanufacture
		repurpose
	Useful application of materials	recycle
LINEAR		recover

Fig. 1: The ten "re-words" continuum of the circular economy. (Source: author's illustration based on Potting et al. 2017)

Fitting the model of a traditional linear economy, we find measures that aim for the useful application of materials, namely recovery and recycling. In this framework, recycling is situated towards the linear end of the continuum (Reike et al. 2018, 257) because existing recycling techniques mostly process materials at a high cost due to the required energy input or expensive technological equipment. Moreover, the resulting secondary materials are predominantly of lower quality, which is why some authors use the term down-cycling (Reike et al. 2018, 256). Given that recycled materials rarely compete directly with primary materials for quality reasons they are likely to be produced in addition, rather than instead of these. Zink and Geyer (2017) have therefore attributed a CE rebound effect to insufficient substitutability. In a truly circular economy, by contrast, recycling would retain the original quality of a product (Potting et al. 2017, 5). Available technologies for plastic recycling are highly controversial with regard to their economic rationale, their sustainability and their health impact. With regard to economics, the cost of recycled plastic is unattractive compared to production from virgin material. The recyclability of plastics varies significantly across product applications: it is relatively high for durable products such as window profiles or pipes (at around 80 percent for window profiles, Rewindo 2020, 3), whereas the recycling of plastic packaging, by contrast, remains comparatively low (at around 40 percent at EU-average, Eurostat 2021). Mechanical recycling, which is the technology most applied and developed, cannot easily be used in the treatment of mixed waste. Chemical recycling has the potential of decomposing mixed waste and complex materials into their virgin substances, yet these technologies are not yet fully developed (Jeswani et al. 2021; World Economic Forum et al. 2016, 47–48). Moreover, since chemical recycling is highly energy intense and thus poses sustainability issues environmental organizations in particular have put into doubt its sustainability (Hann and Connock 2020; Manžuch et al. 2021). Finally, recycled materials may pose health issues in their use stage if these materials contain legacy contaminants. Such legacy contaminants (see also Muncke and Zimmermann in this volume) are chemicals that in the past were added to plastics, but in the meantime have been phased out and banned due to demonstrated health risks. These are, however, still part of secondary material processed for recycling. Policymakers face a tension between the option to discard recycling in cases where legacy substances are involved, and the option that seeks to strike a balance between the environmental benefits of recycling and the management of potential exposure to chemicals. The difficulty of retracing all substances contained in secondary material and the complex assessment of actual exposure to chemicals in the use stage complicate matters further (Friege et al. 2019; European Commission 2020b, 6).

The conceptualization of the intermediary stages that aim at increasing the lifespan of products and their parts includes five categories, namely: repurpose, remanufacture, refurbish, repair, and reuse, in order of increasing ambition. These categories matter to a very different extent depending on the varying lifespan of the products plastics are used in, ranging from packaging and single-use items to long-lasting products such as window profiles or pipes. The largest share of plastic waste, that is packaging waste, will not be suitable for these intermediary "re"-measures, with the exception of reusing packaging items such as plastic bags as waste bags, for example.

Finally, according to the continuum of Potting et al., measures that strive for smarter product use and manufacture are closer to a circular model. Potting et al. define reduce as increased efficiency in product manufacture and in the use of natural resources and materials (Potting et al. 2017, 5). Rethink is about making product use more intensive, with two options of either putting multi-functional products on the market or of sharing products. The most circular measure is to refuse a product, understood as making "products redundant by abandoning its function or by offering the same function with a radically different product" (Potting et al. 2017, 5). Refuse thus may simply mean the restriction of a certain type of product or material, or to replace it with a more sustainable alternative. Policy measures to ban certain plastic materials or applications would fall into this category, as would the replacement of fossil-based plastics by bio-based plastics. Bioplastics have an insignificant market penetration of less than one percent of global plastic production thus far, but this percentage is predicted to grow over the next few years (European Bioplastics 2021). Bioplastics, however, only qualify as a sustainable replacement when they are produced from bio-based feedstock and are biodegradable, as also discussed by Muncke and Zimmermann in this volume.

Analyzing the Policy-Industry Nexus

In recent years we have seen the growth of CE regulatory activities as many governments have gradually intensified their efforts to improve the efficiency of primary resource use, reduce waste production and increase material recovery and reuse (Hartley et al. 2020; Zhu et al. 2019). We also see increasing corporate activity on CE issues with the adoption of industry standards, voluntary agreements, and development of CE business models

(Veleva and Bodkin 2018). Policy measures, including those on plastics, may include a diverse array of initiatives including bans, taxes or voluntary agreements and may target various aspects of the plastics lifecycle such as the reduction of single-use items, increased recycling or more responsible disposal. Moreover, a varying degree of regulatory stringency and leeway for industry is involved.

Typically, the controversy regarding the type of approach and measures to be adopted starts at the agenda-setting and policy formulation stage. There is a rich and growing literature on stakeholder involvement in CE policy frames and discourses (Eckert 2021; Leipold 2021; Leipold and Petit-Boix 2018). This focus is in line with the literature on business power which perceives of corporate discourse as one facet of power (Fuchs 2007, 52–67; Mikler 2018, 35–49). Discursive power plays a central role in shaping perceptions and norms in the agenda-setting phase. In particular, industry will seek to intervene in the policy debate when faced with tangible regulatory threats (Halfteck 2008; Héritier and Eckert 2008) so as to frame the debate, persuade policymakers, and (re)establish its moral legitimacy. An important resource industry draws upon in this discursive battle is its sector-specific expertise (Ambrus et al. 2014; Boswell 2009; Fischer 1990; Fleck et al. 2016; Radaelli 1995; Schrefler 2013), which does not go uncontested (Sending 2015). The power game materializes as a process of generating "counter policy expertise" (Fischer 1990, 28), notably between industry and environmental groups.

It is not only communication and discourse, however, which is strategically mobilized by industry actors. Producing tangible policy output in the form of voluntary standards or self-regulation is another, often highly effective form of alleviating policy pressure, or even of pre-empting looming regulation (Glachant and Finon 2003; Halfteck 2006; 2008; Héritier and Eckert 2008). This rule-setting capacity is another, more structural element of business power (Fuchs 2007, 66). Moreover, industry plays a proactive role in the transition to the CE when it invests in new technologies and drives innovation. Teaming up with industry actors in order to mobilize their input therefore often constitutes a core feature of CE policy measures. It, however, also opens the way for industry to contain the circular policy agenda (Mah 2021).

The Trajectory of EU Environmental Policy on Plastics

Before moving to a more comprehensive CE approach, EU policy measures on plastics were adopted with a differentiated focus on life cycle stages, materials and product applications as a targeted response to policy controversies and public pressure. When addressing environmental issues, the focus typically regarded the waste phase, while health related regulation concerned mostly the use phase. In their contribution to this volume, Muncke and Zimmermann analyze food contact materials, a product group where human health consequences are of particular relevance and subject to comprehensive risk assessment. In what follows, I will examine an industry sector that has attracted attention with regard to both environmental and health-related concerns, and is an illustrative case in preparing the ground for the discussion around the CE and plastics: poly-vinyl-chloride (PVC).

The Debate around PVC

Health-related and environmental issues of PVC have risen to the attention of European policymakers in numerous instances. PVC is a synthetic plastic polymer that, as a rigid material, is widely used in the construction and building sector, while flexible applications, which require the use of plasticizers, are used in a variety of products such as the healthcare sector or toys. Greenpeace campaigned for a total ban of PVC products in Scandinavia, Germany and Austria throughout the 1980s, and from the 1990s onwards also at European level (Interview environmental organization 2006; Interview European Parliament 2005). The environmental organization raised both environmental concerns as to whether sustainably recycling the material was possible, as well as to health problems caused by hazardous and toxic substances in PVC products. Regarding health threats environmentalist groups argued that phthalates, plasticizers frequently used in PVC products, exhibited endocrine disrupting qualities. They targeted a particularly sensitive product group, namely plastic toys intended to be used by babies and small infants (Vogel 2012, 203–204). The ultimate demand of the NGO campaign was to ban PVC products altogether and replace them with more sustainable materials. Relating back to the "re-words" discussed in the previous section, environmental groups thus made a strong case for refuse and challenged the industry's sustainability discourse on recycle.

These campaigns were conducive to policy measures being envisaged and adopted at the European level. A first set of measures focused on health issues and the exposure of babies and small infants in particular. As of 1999 EU policymakers iterated a provisional ban on phthalates in toys in response to campaigning by environmental organizations. As a consequence, the use of six phthalates has been restricted for this product group multiple times: di-iso nonyl phthalate (DINP), di (2-ethylhexyl) phthalate (DEHP), dibutyl phthalate (DBP or DNBP), di-iso-decyl phthalate (DIDP), di-n-octyl phthalate (DNOP) and butylbenzyl phthalate (BBP). These time-limited bans were eventually made permanent when a directive came into force in 2005 (2005/84/EC). Moreover, from the mid-1990s onwards the European Commission sought to generate evidence on the environmental impact of PVC, commissioning several studies (ARGUS 2000; PROGNOS 2000). These studies raised concern as to the use of certain chemicals used in the production of PVC, and to increasing amounts of post-consumer waste. Pressure on the Commission gained momentum in the context of the negotiations of the directive on end-of-life vehicles proposed in 1997. In response to demands by the European Parliament the directive, adopted in 2000, explicitly stated that the Commission would address the environmental impact of PVC in a policy proposal (consideration 12 of directive 53/2000). The Commission published a Green Paper targeting PVC in July 2000 (European Commission 2000). It found that a significant increase in PVC waste was to be expected and that alternatives to landfilling should be developed. More specifically, the proposal envisaged material-specific measures targeting the waste stage of PVC applications.

These proposals were, however, taken off the table because of voluntary industry activity and because of the shift of the policy focus towards the adoption of the EU's regulatory scheme for chemicals, which was finalized in 2008. At this point in time the targeted public campaign came to a halt. Instead, environmental organizations shifted their activities away from PVC in order to focus their attention on new chemicals regulations (Interview European Parliament 2005; Long and Lörinczi 2009). The EU chemical regulation framework for Registration, Evaluation, Authorization and Restriction of Chemicals (REACH) was adopted in 2006 and entered into force in 2007.

Plastics in the Circular Economy

Plastics did, however, return to the spotlight later. In 2016, European environmental organizations joined forces in the alliance "Rethink Plastic" (Rethink Plastic 2018) as part of a global anti-plastic movement (#breakfreefromplastic 2018). The "re-" for reduce and restrict figures prominently in this campaign. Public and NGO pressure has led the EU to adopt a comprehensive policy agenda that covers chemical, product and waste legislation (European Commission 2018a). EU policymakers embarked on adopting a more comprehensive approach in the context of the CE agenda from 2014 onwards. Targeted measures aiming at boosting the circularity of plastics soon became a substantial part of CE actions. Figure 2 gives an overview of policy proposals and legislation relating to plastics in the CE over time (see also Florides and Kramm in this volume).

The policy debate on plastics took off to some degree before the CE agenda did, with the publication of a Green Paper issued in 2013 (European Commission 2013). The proposal describes plastic waste as a "growing problem" (European Commission 2013, 4) for the environment and human health. Moreover, it discusses possible policy options including regulatory and voluntary action on recycling, and the promotion of biodegradable plastics and bio-based plastics. The first tangible measure targeting plastics was proposed in November 2013 and adopted in April 2015, namely a directive (2015/720) aimed at reducing the consumption of lightweight plastic carrier bags.

From December 2014 onwards, the plastics discussion became a cornerstone of the CE agenda, which was launched with the publication of a "zero waste programme for Europe" (European Commission 2014) in September 2014. The 2014 proposal of the outgoing Barroso Commission was however never put into action as the incoming Juncker Commission replaced it with a new action plan "Closing the loop" (European Commission 2015) in December 2015. This case of what in the political science literature has been described as "policy dismantling" by the European Commission caused quite some controversy (Bauer et al. 2012), as it was conceivably a response to industry influence (Confino 2015). The degree of ambition was watered down in the new plan: the 2014 initial CE proposal aimed for a recycling target of 80 percent for packaging by 2030, and a ban on sending recyclable materials, including plastics, to landfill by 2025, with the ambition to "virtually eliminate landfill by 2030" (European

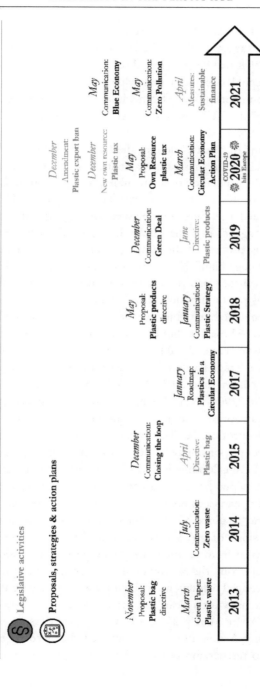

Fig. 2: Timeline EU Policy Measures. (Source: author's illustration)

Commission 2014, 9). The 2015 CE Action plan instead lowered the target for recycling packaging waste to 75 percent and envisioned a ban for landfilling waste that had been collected separately. Based on a comparative analysis Sina Leipold concludes that while the 2014 program sought to transform environmental policy "from within", the 2015 action plan ended up "perpetuating" the EU's discourse of ecological modernization (Leipold 2021).

In the action plan, the EU defines the CE as one "where the value of products, materials and resources is maintained in the economy for as long as possible, and the generation of waste minimised" (European Commission 2015, 2). In light of this, plastics are particularly problematic. The European Commission issued the European plastics strategy as part of the EU's circular economy agenda in January 2018 (European Commission 2018b; for the agenda-setting process of the strategy see Florides and Kramm in this volume). The strategy highlighted various environmental problems such as extremely low levels of plastic recycling and reuse, the emissions of carbon dioxide caused by plastic production and incineration, and marine littering. Moreover, it formulated key objectives to be achieved by 2030: all plastic packaging on the EU market should be either reusable or recyclable; more than half of plastic waste should go to recycling; separate collection of plastics should reach high levels; the consumption of single-use plastic items should be reduced. In June 2019, EU policymakers adopted a directive (directive (EU) 2019/904) to tackle single-use plastics. The directive banned eight product categories, stipulated requirements in terms of binding recycled content, introduced collection targets, and established a scheme of extended producer responsibility (EPR) for a number of single-use products.

The von der Leyen Commission which took office in the second half of 2019 put environmental policy goals center stage with its European Green Deal agenda (European Commission 2019). The Green Deal, amongst other goals, advocated a "resource-efficient and competitive economy where there are no net emissions of greenhouse gases in 2050 and where economic growth is decoupled from resource use" (European Commission 2019, 2). It becomes clear from the wording of the agenda that the European Commission envisaged teaming up with corporate actors with the objective of "mobilising industry for a clean and circular economy" (European Commission 2019, 7). In March 2020, the Commission issued an updated Circular Economy Action Plan which stressed the need to take further action, namely to restrict and reduce microplastics, to facilitate the use of bioplastics, and to increase the uptake of recycled materials (European Commission 2020). More specifically, the Commission announced mandatory requirements for recycled content and waste reduction measures for key products such as packaging, construction materials and vehicles (European Commission 2020, 12). A concrete step was taken in December 2020 when member states agreed on introducing a tax on non-recycled plastic packaging as a new resource category (Article 2 of Council decision (EU, Euratom) 2020/2053).

The CE agenda also feeds into the EU's efforts to foster sustainable finance, given that the transition to a circular economy is one of the environmental goals defined in the green taxonomy (regulation no. 2020/852, article 9). Following adoption of this regulatory framework further detail is specified through several delegated acts. Of relevance for waste treatment and recycling is the delegated regulation adopted in June 2021, which categorizes economic activities according to whether these contribute substantially to climate change mitigation, climate change adaptation, or do not cause significant harm in terms of the environmental objectives defined in the green taxonomy (European Commission 2021a).

The recent European plastics debate, which started off with the issue of marine littering, has come full circle with measures adopted as part of other policy dossiers, including the EU Action Plan on pollution (European Commission 2021c) and the Blue Economy (European Commission 2021b), which target the reduction of plastic litter at sea and microplastics in the environment. Overall, therefore, regulatory pressure on the plastics sector has been heightened during the last decade in the EU. This has resulted in more stringent regulation and more ambitious measures in the transition towards a CE on plastics. There are, however, also signs for a persistent focus on more linear measures such as recycling, which remain problematic for the plastic industry. Finally, the CE framework tends to reinforce the policy-industry nexus, to be discussed in further detail below.

The Evolving Policy-Industry Nexus

Mobilizing the type of industry action needed to move away from the linear model to circularity is a policy challenge, which merits further attention. Revisiting the earlier debate on PVC can help to understand current challenges in the CE in two respects. First, increasing recycling rates was considered the main way to push the sustainability agenda forward back then. This meant that other, more circular options were considered less relevant, a situation we see mirrored in the policy debates around the CE today. Many of the regulatory battles and discursive struggles around PVC recycling seem to be re-emerging, especially where the interface between chemical regulation through REACH and recycling is concerned. The main difference is that environmental issues around recycling are being highlighted now in the context of the current debate on the CE and climate change. Second, voluntary industry action has turned out to be a key component to achieve recycling and other sustainability goals. Used as a preemptive strike in the context of PVC, voluntary agreements have become an implementing tool of the CE policy agenda in the context of the Circular Plastics Alliance.

The Many Challenges of Recycling: Legacy Contaminants and Energy Intensity

How the recycling rate of PVC applications could be increased while also managing the risks around legacy contaminants constitutes a precedent case for the current CE discussion. Indeed the Commission's 2018 non-legislative communication on the interface between chemical, product, and waste legislation repeatedly refers to examples from PVC recycling, namely when it comes to regulating legacy substances, the end-of-waste definition, and the classification of hazardous substances (European Commission 2018a, 4–6). These debates around PVC have proven controversial in many instances. As a rule, the EU chemicals' regulation (REACH) requires the authorization of Substances of Very High Concern (SVHCs). These include lead-based stabilizers and phthalates-based plasticizers (DBP, BBP, DEHP) that in the past have been widely used in PVC products, and thus might reappear once secondary material containing these substances is processed for recycling. As explained earlier, there is a tension between the goal to increase recycling while also managing exposure to chemicals of concern. Whereas REACH requires that recyclers declare the presence of SVHCs once these are present above a certain threshold, exemptions are being granted to facilitate recycling. Such an exemption has been granted to recyclers of DEHP in PVC, very much to the disapproval of members of the European Parliament and environmental groups (European Environmental Bureau 2018; European Parliament 2015; McGrath 2016). Similar controversies reoccur in the context of the CE. Referring back to the "re-words" in the CE continuum, there are significant trade-offs between recycle and restrict: how is it possible to increase recycling, while also taking toxics out? Here the industry seeks to shift the attention away from a discussion about possible restriction of potentially hazardous substances to the overall benefit of achieving higher levels of recycling while making sure there is no exposure to legacy materials for end consumers. From an environmentalist perspective circularity would require to refuse a material such as PVC altogether rather than facilitating its recycling (Interview environmental organization 2017; Interview European Parliament 2017). From an industry perspective, by contrast, regulation, which could be classified as circular, would "give certainty to the industry and investors to boost recycling" (General manager of Vinyl Plus, interviewed by Simon 2016).

Many factors need to be calculated in when assessing the sustainability of existing recycling technologies such as mechanical and chemical recycling from a life cycle perspective (Jeswani et al. 2021). Whether or not chemical recycling ought to be considered sustainable under the newly introduced green taxonomy regulation, for instance, gives rise to controversy. An implementing act that defines economic activities that contribute to climate change mitigation or adaptation (European Commission 2021a) considers the chemical recycling of plastics to be a substantial contribution to climate change mitigation in cases where mechanical recycling is "not technically feasible or economically viable" and the "life-cycle GHG emissions of the manufactured plastic, excluding any calculated credits from the production of fuels, are lower than the life-cycle GHG emissions of the equivalent plastic in primary form manufactured from fossil fuel feedstock" (section 3.17.(b) Annex I). Environmental groups such as Zero Waste objected that the requirements had been watered down throughout the process of adopting relevant policies. According to media coverage (Taylor 2021), the initial proposal had considered chemical recycling only "where mechanical recycling is not possible". Moreover, the requirement that chemical recycled plastic must produce at least 27 percent less life-cycle greenhouse gas emissions than plastic manufactured from virgin feedstock in order to be considered sustainable was removed in the final text. Environmentalists challenge the sustainability of chemical recycling due to its energy intensity, whereas the Commission considers it a promising technology to deal with mixed waste.

The Role of Voluntary Industry Pledges

Voluntary industry action has played an important preventative role in the policy debate around PVC. Back in 2000, at the time when the European Commission proposed measures to target PVC, the European industry chain had already launched a voluntary agreement (Héritier and Eckert 2008). Voluntary action was taken to reduce certain heavy metal stabilizers and to develop mechanical recycling. The PVC industry pursued this path beyond the duration of the first ten-year commitment, and used this format over time to address new policy issues and potential regulatory threats. The most visible quantitative targets of subsequent voluntary agreements concerned recycling. During the first ten-year period the goal was to recycle an additional 200,000 tonnes of available post-consumer PVC waste per year by 2010, followed by a target of 800,000 tonnes per year by 2020 (Vinyl 2010; 2001). For 2000 a study for the European Commission reported 3.6 million tonnes of post-consumer PVC waste, and a recycling rate of three percent (108,000 tonnes), with 82 percent going to landfill and 15 percent to incineration (Brown et al. 2000). Over the course of 2000 to 2015, the PVC industry invested 100 million Euros to develop their recycling capabilities (Simon 2016). In 2021, industry reported to have recycled 6.5 million tonnes of PVC since 2000, with an annual volume of above 700,000 tonnes recycled in 2018, 2019, and 2020. According to industry reporting this amounts to 27.5 percent of the available PVC waste recycled in Europe in 2020 (VinylPlus 2021). Overall, the PVC industry managed to effectively mobilize its regulatory potential in order to avoid their worst-case scenario: the introduction of material-specific regulation, if not a complete ban of PVC. In the ongoing CE discussion, the PVC branch presents itself as a pioneer and role model for the plastics industry as a whole (Interview plastics industry 2019).

The value chain can learn from the PVC example, and it seems that this time around industry responses are very similar. With heightening pressure on plastics, industry has engaged in voluntary action at international and European level over the past decade. In 2011, the plastics value chain signed the Declaration of the Global Plastics Associations for Solutions on Marine

Litter, which included measures to enhance recovery and prevent pellet losses (Marine Litter Solutions 2018). In January 2019, thirty plastic corporations signed the international Alliance to End Plastic Waste, in short AEPW. At the European level industry sought to offset regulatory pressure through concrete policy outputs. PlasticsEurope, the leading European trade association for companies in plastic manufacturing published its initiative "Plastics 2030" the very same day that the Commission came forward with its plastics strategy. The agreement commits to circularity and resource efficiency (PlasticsEurope 2018a). The three priorities of the initiative—to increase reuse and recycling, prevent plastics leakages into the environment, and improve resource efficiency—very much mirror the ongoing EU policy debate discussed in previous sections of this chapter. More specifically, industry voluntarily committed to reuse and recycle 60 percent of plastic packaging by 2030, to increase that share to 70 percent by 2040 (40.8 percent recycled share were reported for 2016, see PlasticsEurope 2018b), and to recycle 50 percent of total plastic waste. These goals are in line with the overall objectives formulated in the plastics strategy, which states that more than half of plastics should go to recycling and that packaging waste should either be reusable or recyclable (European Commission 2018b).

Policymakers have endorsed voluntary industry pledges as part of the implementation process of the CE. The EU's 2018 CE Strategy (annex III) called on stakeholders to make voluntary pledges to use or produce more recycled plastics. To facilitate partnership the Commission initiated the Circular Plastics Alliance (CPA), which brings together organizations along the plastics value chain. Another set of measures to which the European Commission lends support are industry standards governing the quality of sorted plastic waste, as well as of recyclable and recycled plastics. EU policymakers have, however, also introduced more stringent measures despite massive industry lobbying against them. This holds for the single-use plastics directive adopted in 2019. Prior to its adoption PlasticsEurope fought the idea of imposing material-specific restrictions as an unnecessary "shortcut" (PlasticsEurope 2018a). The bioplastics value chain would have wanted to see an exemption for biobased products as a sustainable alternative to conventional single-use items instead (European Bioplastics 2018).

Moreover, a comparative analysis of industry and policymaker discourse points to coevolution and policy congruence (Eckert 2021). The value chain has engaged in concerted action to convince the general public of the benefits of the material. PlasticsEurope is a good example as it promotes the societal and environmental benefits it attributes to its products. The association advocates a "life-cycle driven circular economy" (PlasticsEurope 2019b) meaning that policy measures should not focus solely on "recyclability and reusability at the end-of-life, but also on the benefits for the environment provided during the entire life-cycle." The Covid-19 crisis offered another opportunity for industry to capitalize on related issues: with the unfolding sanitary and health crisis we saw a significant rise in demand for single-use plastic products in various applications such as facemasks and food packaging, as well as for a wide range of medical devices made of or wrapped in plastics, causing a steep increase in plastic waste (Benson et al. 2021; Patrício Silva et al. 2021; Prata et al. 2020). The pandemic also heightened political pressure to act and strive for a "green" exit from the crisis.

Conclusions

This chapter, written from a political science perspective, contributes to the interdisciplinary debate presented in this edited volume with a conceptual discussion of the CE and a focus on the evolving policy approach at the European level and the role of industry therein. Conceptualizing the CE as a continuum of more as well as less ambitious policy measures ranging from those compatible with a linear economic model, as well as those aimed at a truly circular model, illustrates that the CE agenda opens up the space for multiple policy options. The conceptual discussion shows that the CE agenda can still be mostly aligned with positions biased towards safeguarding the status quo. This holds true empirically when we consider the recent debate on plastics in the CE at European level: although pressure to act has grown following a pronounced degree of public and political mobilization, industry was in a position to contain the debate to some degree and claim ownership of the CE through self-regulatory action. In short, with the CE agenda, we see more of the same and not the transformative change, which is potentially needed to embark on a new economic model.

The lack of transformation becomes obvious when considering earlier policy controversies and deliberation on PVC in the late 1990s and early 2000s. Attempts to ban the use of PVC failed despite pronounced public and political mobilization. Instead, industry engaged in a pre-emptive strike through voluntary action, and in the course of the process sought to reframe its role as one of being a sustainability pioneer. Moreover, many regulatory and policy issues are currently re-emerging on issues such as legacy materials in the recycling process. A closer look at the unfolding CE agenda of the EU shows that the bulk of policy measures adopted are more in line with a linear economy, rather than with true circularity. Voluntary commitments by industry have put recycling activities front and center, and have also provided corporate voices with a platform to communicate their preferred policy choices, very similar to what we have seen previously in the course of the PVC debate.

Rather than the CE agenda, it may well be the global climate ambitions that will ultimately challenge the status quo of plastic production and consumption (Mah 2021) in the plasticene (Ross 2018). Although we have seen a degree of product innovation through bio-based and biodegradable plastics, the bulk of industry activity around the globe remains fossil-based. Plastic production will, for the foreseeable future, remain a key branch of the petrochemical industry, and even gain in relative importance: with the growing ambition to decarbonize the transport sector it is forecasted to become the largest source of growth in oil use in the coming decade. According to data provided by the International Energy Agency this would be the case even in a scenario where global recycling rates for plastics were to double, due to continued demand growth originating in Asia in particular (International Energy Agency 2021; Simon 2019). If global society would, by contrast, depart from the path of ever-increasing production and consumption, such de-growth would mean the demise of the oil sector (Carbon Tracker 2020). In short: decarbonizing the economy would mean putting an end to the "plastic age" in which we currently live.

References

#breakfreefromplastic (2022). Break free from plastic global movement. 07.03.2022 https://www.breakfreefromplastic.org/about/

Ambrus, Monika, Karin Arts, Ellen Hey, and Helena Raulus (2014). *The role of 'experts' in international and European decision-making processes: Advisors, decision makers or irrelevant actors?* Cambridge: Cambridge University Press.

ARGUS (2000). *The behaviour of PVC in landfill.* Final Report February 2000. Brussels: European Commission. 03.03.2022 https://edz.bib.uni-mannheim.de/www-edz/pdf/2000/pvclandfill.pdf.

Bauer, Michael W., Andrew Jordan, Christoffer Green-Pederson, and Adrienne Héritier (eds.). (2012). *Dismantling public policy: Preferences, strategies, and effects.* Oxford: Oxford University Press.

Benson, Nsikak U., David E. Bassey, and Thavamani Palanisami. (2021). COVID pollution: impact of COVID-19 pandemic on global plastic waste footprint. *Heliyon*, 7 (2), 1–8.

Blomsma, Fenna, and Geraldine Brennan. (2017). The emergence of circular economy: A new framing around prolonging resource productivity. *Journal of Industrial Ecology*, 21 (3), 603–614.

Boswell, Christina (2009). *The political uses of expert knowledge: Immigration policy and social research.* Cambridge: Cambridge University Press.

Brown, Keith A., M R Holland, R A Boyd, S Thresh, H Jones and S M Ogilvie (2000). *Economic Evaluation of PVC Waste Management.* A report produced for European Commission Environmental Directorat. Oxfordshire: AEA Technology. 03.03.2022 https://ec.europa.eu/environment/enveco/waste/index.htm.

Carbon Tracker (2020, 4 September). The future's not in plastics: Why plastics demand won't rescue the oil sector. 03.03.2022 https://carbontracker.org/reports/the-futures-not-in-plastics/.

Confino, Jo (2015, 3 February). Future of Europe's circular economy mired in controversy. 03.03.2022 https://www.theguardian.com/sustainable-business/2015/feb/03/architect-europe-circular-economy-strategy-lambasts-successors.

Eckert, Sandra (2019). *Corporate power and regulation. Consumers and the environment in the European Union.* London: Palgrave.

Eckert, Sandra (2021). Varieties of framing the circular economy and the bioeconomy: unpacking business interests in European policymaking. *Journal of Environmental Policy & Planning*, 23 (2), 1–13.

European Environmental Bureau (2018, 6 July). 55 NGOs ask Commission to reject authorisation of hazardous DEHP in PVC plastic. 03.03.2022 https://eeb.org/library/55-ngos-ask-commission-to-reject-authorisation-of-hazardous-dehp-in-pvc-plastic/.

European Bioplastics (2018, 20 December). Single-use plastics directive fails to acknowledge potential of biodegradable plastics. 03.03.2022 https://www.european-bioplastics.org/single-use-plastics-directive-fails-to-acknowledge-potential-of-biodegradable-plastics/.

European Bioplastics (2021). Bioplastics market development Update 2021. 03.03.2022 https://docs.european-bioplastics.org/publications/market_data/2021/Report_Bioplastics_Market_Data_2021_short_version.pdf.

European Commission (2000). Green Paper. Environmental issues of PVC. COM (2000) 469 final. 03.03.2022 https://eur-lex.europa.eu/legal-content/EN/TXT/?uri=celex%3A52000DC0469.

European Commission (2013). Green Paper. On a European Strategy on Plastic Waste in the Environment. COM/2013/0123 final. 03.03.2022 https://eur-lex.europa.eu/legal-content/EN/TXT/?uri=CELEX:52013DC0123.
European Commission (2014). Communication from the Commission to the European Parliament, the Council, the European Economic and Social Committee and the Committee of the Regions. Towards a circular economy: A zero waste programme for Europe. COM (2014) 398 final. 03.03.2022 https://eur-lex.europa.eu/legal-content/EN/TXT/?uri=celex%3A52014DC0398.
European Commission (2015). Communication from the Commission to the European Parliament, the Council, the European Economic and Social Committee and the Committee of the Regions. Closing the loop – An EU action plan for the circular economy. COM (2015) 614 final. 03.03.2022. https://eur-lex.europa.eu/legal-content/EN/TXT/?uri=CELEX:52015DC0614.
European Commission (2018a). Communication from the Commission to the European Parliament, the Council, the European Economic and Social Committee and the Committee of the Regions. On the implementation of the circular economy package: options to address the interface between chemical, product and waste legislations. COM (2018) 32 final. 03.03.2022 https://eur-lex.europa.eu/legal-content/en/ALL/?uri=CELEX:52018DC0032.
European Commission (2018b). Communication from the Commission to the European Parliament, the Council, the European Economic and Social Committee and the Committee of the Regions. A European strategy for plastics in a circular economy. COM (2018) 28 final. 03.03.2022 https://eur-lex.europa.eu/legal-content/EN/TXT/?uri=COM:2018:28:FIN.
European Commission (2019). Communication from the Commission to the European Parliament, the European Council, the Council, the European Economic and Social Committee and the Committee of the Regions. The European Green Deal. COM (2019) 640 final. 03.03.2022 https://eur-lex.europa.eu/legal-content/EN/TXT/?uri=COM:2019:640:FIN.
European Commission (2020a). Communication from the Commission to the European Parliament, the European Council, the Council, the European Economic and Social Committee and the Committee of the Regions. A new circular economy action plan. For a cleaner and more competitive Europe. COM (2020) 98 final. 03.03.2022 https://eur-lex.europa.eu/legal-content/EN/TXT/?uri=CELEX:52020DC0098.
European Commission (2020b). Communication from the Commission to the European Parliament, the Council, the European Economic and Social Committee and the Committee of the Regions. Chemicals Strategy for Sustainability. Towards a toxic-free environment. COM (2020) 667 final. 03.03.2022 https://eur-lex.europa.eu/legal-content/EN/TXT/?uri=COM%3A2020%3A667%3AFIN.
European Commission (2021a). Commission Delegated Regulation (EU) of 4.6.2021 supplementing Regulation (EU) 2020/852 of the European Parliament

and of the Council by establishing the technical screening criteria for determining the conditions under which an economic activity qualifies as contributing substantially to climate change mitigation or climate change adaptation and for determining whether that economic activity causes no significant harm to any of the other environmental objectives. C (2021) 2800 final. 03.03.2022 https://eur-lex.europa.eu/legal-content/EN/TXT/?uri=PI_COM:C(2021)2800.

European Commission (2021b). Communication from the Commission to the European Parliament, the Council, the European Economic and Social Committee and the Committee of the Regions. On a new approach for a sustainable blue economy in the EU. Transforming the EU's blue economy for a sustainable future. COM (2021) 240 final. 03.03.2022 https://eur-lex.europa.eu/legal-content/EN/TXT/?uri=COM:2021:240:FIN.

European Commission (2021c). Communication from the Commission to the European Parliament, the Council, the European Economic and Social Committee and the Committee of the Regions. Pathway to a Healthy Planet for All. EU Action Plan: 'Towards zero pollution for air, water and soil'. COM (2021) 400 final. 03.03.2022 https://eur-lex.europa.eu/legal-content/EN/TXT/?uri=CELEX:52021DC0400.

European Parliament (2015, 25 November). Don't allow recycling of plastics that contain toxic phthalate DEHP, warn MEPs. 20151120IPR03616. 03.03.2022 https://www.europarl.europa.eu/news/en/press-room/20151120IPR03616/don-t-allow-recycling-of-plastics-that-contain-toxic-phthalate-dehp-warn-meps.

Eurostat. (2021a, 27 October). EU recycled 41% of plastic packaging waste in 2019. 03.03.2022 https://ec.europa.eu/eurostat/web/products-eurostat-news/-/ddn-20211027-2.

Fischer, Frank (1990). *Technocracy and the politics of expertise*. Newburz Park, CA: Sage Publications.

Fleck, James, Wendy Faulkner, and Robin Williams (2016). *Exploring expertise: Issues and perspectives*. London: Palgrave Macmillan.

Florides, Paula and Johanna Kramm (2022). Explaining agenda-setting of the European plastics strategy. A multiple streams analysis. In Johanna Kramm and Carolin Völker (eds.). *Living in the plastic age. Perspectives from humanities, social sciences and environmental sciences*. Frankfurt, New York: Campus.

Friege, Henning, Beate Kummer, Klaus Günter Steinhäuser, Joachim Wuttke, and Barbara Zeschmar-Lahl (2019). How should we deal with the interfaces between chemicals, product and waste legislation? *Environmental Sciences Europe*, 31 (51) 1–18.

Fuchs, Doris (2007). *Business Power in Global Governance*. Boulder: Lynne Rienner.

Geissdoerfer, Martin, Paulo Savaget, Nancy M. P. Bocken, and Erik J. Hultink (2017). The circular economy – A new sustainability paradigm? *Journal of Cleaner Production*, 143, 757–768.

Glachant, Jean-Michel, and Dominique Finon (eds.). (2003). *Competition in European electricity markets. A cross-country comparison.* Cheltenham, Northhampton: Edward Elgar.

Halfteck, Guy (2006). *A theory of legislative threats.* Bepress Legal Series. Working Paper, 1122. 03.03.2022 http://law.bepress.com/expresso/eps/1122/.

Halfteck, Guy (2008). Legislative threats. *Stanford Law Review*, 61 (3), 629–710.

Hartley, Kris, Ralf van Santen, and Julian Kirchherr (2020). Policies for transitioning towards a circular economy: Expectations from the European Union (EU). *Resources, Conservation and Recycling*, 155, 104634.

Héritier, Adrienne, and Sandra Eckert (2008). New modes of governance in the shadow of hierarchy: Self-regulation by industry in Europe. *Journal of Public Policy*, 28 (1), 113–138.

Hann, Simon, and Toby Connock (2020). *Chemical recycling: State of play.* Report for CHEM Trust. 03.03.2022 https://www.eunomia.co.uk/reports-tools/final-report-chemical-recycling-state-of-play/.

International Energy Agency (2021). World Energy Outlook 2021. 03.03.2022 https://www.iea.org/reports/world-energy-outlook-2021.

Interview environmental organization (2006). Head of Office. Brussels: Greenpeace European Unit, 22.02.2006.

Interview environmental organization (2017). Executive director. London: CHEM Trust, 17.07.2017.

Interview European Parliament (2005). Policy officer. Brussels: Greens/EFA, 23.11.2005.

Interview European Parliament (2017). Policy officer. Brussels: Greens/EFA, 12.04.2017.

Interview plastics industry (2019). Technical and environmental affairs manager, public affairs senior manager. Brussels: ECVM, 11.01.2019.

Jeswani, Harish, Christian Krüger, Manfred Russ, Maike Horlacher, Florian Antony, Simon Hann, and Adisa Azapagic (2021). Life cycle environmental impacts of chemical recycling via pyrolysis of mixed plastic waste in comparison with mechanical recycling and energy recovery. *Science of the Total Environment*, 769, 144483.

Jordan, Andrew, and Camilla Adelle (eds.). (2013). *Environmental Policy in the EU: Actors, Institutions and Processes.* Ed. third edition. London, New York: Routledge.

Leipold, Sina (2021). Transforming ecological modernization 'from within' or perpetuating it? The circular economy as EU environmental policy narrative. *Environmental Politics*, 30 (6), 1045–1067.

Leipold, Sina, and Anna Petit-Boix (2018). The circular economy and the bio-based sector – Perspectives of European and German stakeholders. *Journal of Cleaner Production*, 201, 1125–1137.

Lee, Roh Pin, Manja Tschoepe, & Raoul Voss (2021). Perception of chemical recycling and its role in the transition towards a circular carbon economy: A case study in Germany. *Waste Management*, 125, 280–292.

Long, Tony, and Larisa Lörinczi (2009). Business Lobbying in the European Union. In David Coen and Jeremy Richardson (eds.). *Lobbying the European Union: Institutions, actors, and issues*, 169–185. New York: Oxford University Press.

Mah, Alice (2021). Future-proofing capitalism: The paradox of the circular economy for plastics. *Global Environmental Politics*, 21 (2), 121–142.

Manžuch, Zinaida, Rūta Akelytė, Marco Camboni, and David Carlander (2021). Chemical recycling of polymeric materials from waste in the circular economy. RPA Europe. Final report prepared for the European Chemicals Agency. 03.03.2022 https://echa.europa.eu/documents/10162/1459379/chem_recycling_final_report_en.pdf/887c4182-8327-e197-0bc4-17a5d608de6e?t=1636708465520.

Marine Litter Solutions (2018). The Delcaration of the Global Plastics Associations for Solutions on Marine Litter. 4th Progress Report. March 2018. 05.04.2018 https://www.marinelittersolutions.com/wp-content/uploads/2018/04/Marine-Litter-Report-2018.pdf.

McDowall, Will, Yong Geng, Beijia Huang, Eva Barteková, Raimund Bleischwitz, Serdar Türkeli, René Kemp, and Teresa Doménech (2017). Circular economy policies in China and Europe. *Journal of Industrial Ecology*, 21(3), 651–661.

McGrath, Meredith (2016, 21 April). EU approves use of recycled plastics containing DEHP. 03.03.2022 https://www.reuters.com/article/us-europe-regulations-plastics-idUSKCN0XI29T.

Mikler, John (2018). *The political power of global corporations*. Cambridge: Polity Press.

Muncke, Jane, and Lisa Zimmermann (2022). Chemicals in plastics: Risk assessment, human health consequences, and political regulation. In Johanna Kramm and Carolin Völker (eds.). *Living in the plastic age. Perspectives from humanities, social sciences and environmental sciences*. Frankfurt, New York: Campus.

Murray, Alan, Keith Skene, and Kathryn Haynes (2017). The circular economy: An interdisciplinary exploration of the concept and application in a global context. *Journal of Business Ethics*, 140 (3), 369–380.

Patrício Silva, Ana L., Joana C. Prata, Tony R. Walker, Armando C. Duarte, Wei Ouyang, Damià Barcelò, and Teresa Rocha-Santos (2021). Increased plastic pollution due to COVID-19 pandemic: Challenges and recommendations. *Chemical Engineering Journal*, 405, 1266–1283.

PlasticsEurope (2018a). Plastics 2030. PlasticsEurope's voluntary committment to increasing circularity and resource efficiency. 03.03.2022 https://plasticseurope.org/knowledge-hub/plastics-2030-plasticseuropes-voluntary-commitment-to-increasing-circularity-and-resource-efficiency/.

PlasticsEurope (2018b). Plastics - the Facts 2017. An analysis of European plastics production, demand and waste data. Brussels, Wemmel: Association for Plastics Manufacturers, European Association of Plastics Recycling. 03.03.2022 https://plasticseurope.org/wp-content/uploads/2021/10/2017-Plastics-the-facts.pdf.

PlasticsEurope (2019a). Plastics – the Facts 2019. An analysis of European plastics production, demand and waste data. Brussels, Wemmel: Association for Plastics Manufacturers, European Association of Plastics Recycling. 03.03.2022 https://plasticseurope.org/wp-content/uploads/2021/10/2019-Plastics-the-facts.pdf.

PlasticsEurope (2019b). Towards a life-cycle driven circular economy. PlasticsEurope. 03.03.2022 https://plasticseurope.org/knowledge-hub/towards-a-life-cycle-driven-circular-economy/.

Potting, José, Marko Hekkert, Ernst Worrell, and Aldert Hanemaaijer (2017). *Circular economy: Measuring innovation in the product chain. Policy report.* Utrecht: PBL Netherlands Environmental Assessment Agency. 03.03.2022 https://www.pbl.nl/en/publications/circular-economy-measuring-innovation-in-product-chains.

Prata, Joana C., Ana L. P. Silva, Tony R. Walker, Armando C. Duarte, and Teresa Rocha-Santos (2020). COVID-19 pandemic repercussions on the use and management of plastics. *Environmental Science and Technology*, 54 (13), 7760–7765.

PROGNOS (2000). *Mechanical Recycling of PVC Wastes.* Bruessels: European Commission. 03.03.2022 https://www.environmental-expert.com/articles/mechanical-recycling-of-pvc-wastes-1585.

Radaelli, Claudio (1995). The role of knowledge in the policy process. *Journal of European Public Policy*, 2 (2), 159–183.

Reike, Denise, Walter J. V. Vermeulen, and Sjors Witjes (2018). The circular economy: New or refurbished as CE 3.0? — Exploring controversies in the conceptualization of the circular economy through a focus on history and resource value retention options. *Resources, Conservation and Recycling*, 135, 246–264.

Rethink Plastic (2018). Rethink plastic alliance of leading European NGOs. 04.08.2018 http://www.rethinkplasticalliance.eu/.

Rewindo (2020). Kunststofffenster-Recycling in Zahlen 2020. 03.03.2022 https://rewindo.de/download/2118/

Ross, Nancy (2018). The "plasticene" epoch? *Elements*, 14 (5), 291.

Schrefler, Lorina S. (2013). *Economic knowledge in regulation: The use of expertise by independent agencies.* Colchester: ECPR Press.

Sending, Ole J. (2015). *The politics of expertise: Competing for authority in global governance.* Michigan: University of Michigan Press.

Simon, Frédéric (2016). PVC boss: We will not change our reputation in one day. Euractiv. 03.03.2022 https://www.euractiv.com/section/sustainable-dev/interview/vinyplus-boss-we-will-not-change-our-reputation-in-one-day/.

Simon, Frédéric (2019). How Europe's war on plastics is affecting petrochemicals. Euractiv. https://www.euractiv.com/section/energy/news/how-europes-war-on-plastics-is-affecting-petrochemicals/.

Taylor, Kyra (2021). Last minute EU taxonomy changes water down sustainability criteria for waste, NGOs say. 03.03.2022 https://www.euractiv.com/

section/circular-materials/news/last-minute-eu-taxonomy-changes-water-down-sustainability-criteria-for-waste-ngos-say/.
Töller, Annette E. (2010). Measuring and comparing the Europeanization of national legislation: A research note. *Journal of Common Market Studies*, 48 (2), 417–444.
Veleva, Vesela, and Gavin Bodkin (2018). Corporate-entrepreneur collaborations to advance a circular economy. *Journal of Cleaner Production*, 188, 20–37.
Vinyl 2010 (2001). The voluntary commitment of the European PVC industry, October 2001. Brussels: Vinyl 2010.
VinylPlus (2021). Reporting on 2020 Activities and summarising the key achievement of the past 10 years. 03.03.2022 https://www.vinylplus.de/wp-content/uploads/2021/06/VinylPlus-Progress-Report-2021_WEB.pdf.
Vogel, David (2012). *The politics of precaution: Regulating health, safety and environmental risks in Europe and the United States*. Princeton: Princeton University Press.
World Economic Forum, Ellen MacArthur Foundation, and McKinsey & Company (2016). The New Plastics Economy – Rethinking the future of plastics. 03.03.2022 https://ellenmacarthurfoundation.org/the-new-plastics-economy-rethinking-the-future-of-plastics.
Zhu, Junming, Chengming Fan, Haijia Shi, and Lei Shi (2019). Efforts for a circular economy in China: A comprehensive review of policies. *Journal of Industrial Ecology*, 23 (1), 110–118.
Zink, Trevor, and Roland Geyer (2017). Circular economy rebound. *Journal of Industrial Ecology*, 21 (3), 593–602.

Plastivores and the Persistence of Synthetic Futures

Kim De Wolff

Introduction

A white-clothed table is set with a fork, a knife, and anticipation. The meal arrives and you squint to confirm that the plump red cherry tomatoes and lush sprigs of green parsley adorn a bed of, yes, crumpled grey plastic bag. A woman in a simple black dress enters the frame, seats herself at the table, and as she arranges a cloth napkin in her lap, your eyes widen at her seeming indifference to the meal's synthetic contents. Without hesitation, she raises a large forkful of plastic salad to her mouth, and your throat tightens, stomach clenches, you close your eyes in disgust. The audio is as sparse as the place setting: the clinking of cutlery against ceramic and the grating crinkle of plastic being methodically chewed, bite after bite, until the last morsel is washed down with a casual gulp of wine. Only then does the woman meet your gaze, and a message appears in bold white capital letters: "You wouldn't eat plastic bags. So why should marine life?" Images of turtles doing just that dominate the screen. The final message from this 2016 Green Peace Australia video is clear: take plastic bags off the menu.

The affective power of the "Take Plastics Off the Menu" campaign rests on a deeply embodied understanding that plastic is not food. It works from the assumption that the not-food-ness of plastic is common knowledge to the point that no further explanation is needed, the visceral reaction of indignation is both expected and sufficient. Even as viewers attempt to explain away what they are seeing—the plastic is in her cheek/she spits it out/it's not really plastic—the bodily response defies logical excuses. With cringes and gags, the viewer becomes a participant in a powerful narrative where plastic is an unequivocally "bad" actor that needs to be kept out of food chains for humans and non-humans alike. It needs to be cleaned up, if not banned. As in many campaigns that preceded it, the plastic bag is "an object of moral condemnation," that "disrupts the binaries between pure and contaminated" (Hawkins 2006, 22). This is a story played out again and again: in how-to guides for plastic-free living; in children's science books warning of the health impacts of plastic in food; in ocean clean-up campaigns focused on marine life ingesting plastic, and the humans that eat them in turn.[1] The dominant practices of plastic pollution activism, journalistic and scientific writing, all materialize a foundational assumption that synthetic plastics and living bodies should not mix (De Wolff 2017).

However, a parallel narrative is emerging, one where eating plastic is desirable, even celebrated. Since 2016, a spate of headlines has announced the discovery of a growing menagerie of organisms capable of not only ingesting, but digesting some types of synthetic plastics. For these "plastivores," some kinds of plastics are food—a source of energy and nutrients that can be metabolized to fuel the basic processes of life. Media coverage was quick to speculate about bacteria and insect larvae's potential for "taking a bite out of the recycling problem" (Cornwall 2021), offering a "viable solution to the plastic problem" (Mulhern 2021), or even "saving the planet" (Ap 2016). These plastic-eaters clearly have not only plastic, but aspirations for a synthetic circular economy caught up with the capacity of their guts. Plastic on the menu turns from a problem to a panacea: eating plastic just might be the remedy for planetary plastic woes.

This paper explores how plastivores are gnawing away the foundations of plastic-life relations and reconfiguring the possibilities for living responsibly in the Plastic Age. Growing knowledge that synthetic plastics might be digestible challenges cultural assumptions about what makes plastic a "bad" substance as seen in the common tropes of so many anti-plastics campaigns. Plastic may not belong on human dinner plates, but if other species can incorporate it into bodily processes, then maybe it isn't so unnatural or even harmful after all. Moreover, being metabolized potentially undermines plastic's prodigal endurance, as its uncanny capacity to last seemingly forever is enfolded back into more familiar lifecycles of decay and regeneration. If

---

[1] See for example, Michael San Clements (2014) Plastic Purge: How to Use Less Plastic, Eat Better, Keep Toxins Out of Your Body, and Help Save the Sea Turtles!; Maria Simon (2020) You Are Eating Plastic!; Danielle Smith-Llera (2020) You Are Eating Plastic Every Day; the plastic-bag sushi in the Surfrider "What Goes in the Ocean Goes in You" campaign; and a plethora of stock photos where models awkwardly "eat" bottles and bags.

the material can now be properly biodegraded, then why worry about producing and consuming less of it? Is there even a plastic problem at all? Or so goes increasingly popular reasoning.

For many science and technology studies scholars, however, plastic was never inherently "bad" in the first place. Instead, these scholars have focused on how a multiplicity of plastics are part of diverse practices where the ethical and political emerge in ongoing relationships, rather than being fixed in advance. Here "plastics set in motion relations between things that become sites of responsibility and effect" (Gabrys et al. 2013, 5). This is not to say plastic is good, but to make space for ambiguity beyond good/bad material binaries, and insist on understanding what plastics—plural—do in specific situations. Furthermore, taking relationships as the basis of analysis moves conversations about what to do away from an ethic of individual consumer choice, and toward conversations about collective responsibilities and socio-economic structures of extraction, production, and waste (Liboiron 2021; Shotwell 2016). From this relational perspective, where plastivores are neither inherently good or bad, different questions arise. I ask instead, what kinds of relationships are emerging between plastic-eaters, technoscience, and petro-capitalism?

This chapter begins with a discussion of the plastivore concept, and the intricacies of what exactly counts as eating plastic. The term first emerges to describe ingestion, plastic entering bodies through the same pathways as food. As plastivory comes to be associated with organisms that can biodegrade plastics, however, it simultaneously becomes entangled with specific visions for plastics recycling. Biodegradability, which conjures images of plastic decaying back into earth, might seem be to be the opposite of recycling which ideally implies seamless continued economic circulation of materials (for a critique on plastics recycling see Eckert in this volume). However, with the extraction and bioengineering of specific enzymes from plastic-eating organisms, the goal becomes breaking plastics into constituent building blocks that can then be reprocessed into "new" plastics rather than compost.

I further explore this tension between degradation and circularity by reading plastivore science against feminist and queer theoretical engagements with plastics and toxicity. Where speculation about plastic-eating lifeforms has provided a generative substrate for creatively imaging new kinds of plastic-life relations in fiction, art, and theory, recent scientific confirmation that organisms are eating plastic seem to point to something at once very different, and at the same time, disconcertingly familiar: plastivores are emerging as a biotechnological fix that wrangles stubborn synthetics back into circuits of capitalist production. Plastic-eating is actively produced as "good" as it is coopted into extractive capitalism and the service of a synthetic circular economy. I show how plastivores are a "solution" only when the problem is narrowly framed as a stubborn physical quality of synthetic materials in isolation; they do not address broader relations of domination and control that assume materials exist to be manipulated for the benefit of (some) humans (Davis 2022).

Plastivores can help instigate a radical shift in understandings of how to live with plastic, but this involves actively resisting the allure of material circularity in order to challenge the endurance of interconnected social and political relations. I add to this project by focusing on the relations of plastic persistence across scales, from the chemical-microbial to the technoscientific-economic.

Persistent organic pollutants that refuse to conform to neat scientific narratives about efficient recycling help illuminate the persistence of a petro-capitalist status quo where biotechnological fixes are mobilized to ensure continued extraction, production, and circulation with their associated harms. A focus on persistence brings attention to the connections between plastic's material endurance and the future of petro-cultures. Persistence shows how the emergence of plastivores alone does not undermine the imbricated if always unequal accretions of wealth and harm associated with plastic's accumulation (Gabrys et al. 2013). Persistence counters assumptions of limitless plasticity by stubbornly tracing the enduring relations and facing them head on, potential provoking systemic change.

The Evolution of Plastic-Eaters

Plastivore is a neologism describing new kinds of biological relationships with plastics. Like other "vores," the term is rooted in the Latin vorare—devouring—which denotes feeding on a specific kind of food, in this case plastic. The word plastivore itself is a declaration that plastic is food and at the same time, that plastic is categorically distinct from other substances that are eaten. As with many new terms, not everyone uses it in the same way. Whether or not a plastic-life relationship qualifies as plastivorous, differs depending on what exactly eating entails, which substances qualify as plastic, and whether intention matters. Most importantly, plastivores emerge with visions of desirable plastic-life relations already attached. From science-oriented accounts, ideal relations appear to variously involve encouraging new associations or maintaining material separations, depending on whether plastic is understood to fuel processes of life or seen as a hindrance to them.

In the first known use of the term "plastivore" in print, eating is equated with ingestion. As identified by Haram et al. (2020), this occurs in Donovan Hohn's (2011) Moby Duck, a nonfiction account of the author's pursuit of plastic bath toys mysteriously washing up on beaches around the world. With evocative prose, Hohn makes a single use of the term to describe Laysan albatross, the large seabirds infamous for having bellies full of plastic, as "probably the most voracious plastivore on the planet" (Hohn 2011, 72). For Hohn, plastic is a catch-all category of synthetic substances which enter animal bodies when mistaken for food. As with the Greenpeace Australia campaign, eating in this case, is a decidedly bad relationship to have with plastic; at best, it is unproductive because plastic cannot be digested, and worse, it may be deadly, although establishing causality between plastic ingestion and albatross chick deaths to scientific standards remains elusive (Liboiron 2021, 104–105).

As plastic pollution, especially ocean plastic pollution, has increasingly become a focus of scientific study, a similar sense of plastic eating as ingestion has since been applied retroactively. In "Zooplankton Plastivory," Xabier Irigoien (2018) compiles decades of existing research on plastic ingestion, tracing origins to 1970s laboratory studies where copepods were intentionally offered plastic pellets in order to understand whether or not the tiny crustaceans had agency in food selection. Here, and in studies of plankton feeding behavior that followed, microplastic pellets were described as "inert particles" (Bundy and Henry 2002), chemically and morally decoupled from pollution as either litter or toxin. The agential capacity of plankton rather than of plastic is what was in question, with the assumption that these organisms would not ingest plastic if they indeed had the ability to choose since plastic was decidedly not food as it could not fuel the processes of life.[2]

---

[2] If eating is understood to include unintentional ingestion, then we humans may all be plastivores. Despite the cultural convictions that humans should not eat plastic, study after study has identified the myriad paths tiny plastic particles take into and through human bodies. This includes eating other beings that have themselves ingested plastic, like

For others, plastic-eating involves relationships beyond mere swallowing, whether passively or by mistake. In their "Plasticene Lexicon," Haram et al. (2020) define plastivores as "any organism that ingests, processes, and regurgitates or defecates plastic materials." Here, eating is part of a longer trajectory of relationships where consuming plastic may also involve its transformation. Though there was speculation that bacteria inhabiting tiny pits on floating ocean plastics had themselves made the divots by eating the plastic (Zettler et al. 2013), it is only very recently that scientific studies have confirmed that life forms degrade plastics (rather than sun exposure or mechanical forces like wave action). Researchers have identified, for example, marine polychaete worms as themselves microplastic "producers" that actively fragment larger plastic objects by biting off chunks and pushing them through their digestive tracks (Jang et al. 2018). Under laboratory conditions, "a single polychaete can produce hundreds of thousands of microplastics a year" (Jang et al. 2018, 365). Again, plastic eating is problematic, in this case as non-humans become active agents in a process of creating rather than mitigating ocean plastic pollution.

While some species gnaw plastic into smaller pieces, only for select organisms does plastic-eating involve relationships of metabolism—breaking down synthetic chemical bonds at the molecular level and transferring energy and nutrients necessary for the processes of life. These species are mostly bacteria, though their ranks have grown to include fungi and insect larvae. The first confirmed report published in the western scientific literature comes from a study of polyethylene terephthalate (PET) plastic-contaminated soil and sludge from a Japanese recycling facility. Yoshida et al. (2016) identified a previously unknown species of bacteria that was not just eating, but living with and from plastic. Laboratory isolation confirmed that the bacteria use PET as "a major carbon source for growth"

---

seafood, but also less expected source such as bottled water and beer (Campanale et al. 2020). And yes, there is plastic in our poop. Microplastic particles are so adept at permeating body borders, that plastic has even been found in the first meconium excretions of newborn humans who, having not yet eaten through their own mouths, have absorbed it through the placenta (Simon 2021). Becoming plastivores because of the inability to avoid eating plastic complicates the dominant narratives surrounding the ethics of eating which tend to be dominated both by discussions of meat consumption and by assumptions of atomistic individual subjects making conscious decisions about what enters their bodies or not. In decoupling eating from intention, plastivory requires an approach beyond individual choice frameworks—as seen with feminist approaches to food (Shotwell 2018). If eating entails metabolism, however, then humans are arguably plastivores only metaphorically.

(Yoshida et al. 2016, 1196). While the authors are describing relationships of flourishing microbial bodies, as a humanities scholar I cannot help but read this statement as a metaphor for capitalism where even contamination has become the substrate for yet more production. As the article swiftly moves from the study of "consortiums" of materials and organisms, to isolating specific enzymes (a type of protein that speeds up bodily processes including digestion in all living things) bacteria produce, so too does it explicitly identify a "platform for biological recycling of PET waste products" (Yoshida et al. 2016, 1196). The first scientifically certain evidence of microorganisms making their living with plastics is from the very start entwined with visions not only of recycling, but also of shifting plastics recycling toward metabolic life processes.

In the subsequent scientific articles and popular news coverage of them that follow, coupling plastivores with recycling is the norm not the exception. With an understanding that plastic pollution is a waste management problem for which recycling is the unquestioned solution, plastic-degrading microbes emerge as and are articulated again and again as a biotechnical fix caught up with dreams for a circular plastic economy (Carpenter 2021; Cornwall 2021). A recent large-scale study identifying a potential 30,000 existing plastic-degrading enzymes lauds microorganisms' "great potential to revolutionize the management of global plastic waste" (Zrimec et al. 2021, 9). The correlation of plastivorous abundance with the spatial distribution of pollution becomes evidence that earth is adapting to plastic pollution and—at the same time—that this "natural" process must be efficiently harnessed for sustainable management. Though evolution appears to be so very conveniently aligning with petro-capitalist human plans, microbial "carbon workers" (Gabrys 2013), are deemed to work too slowly without technoscientific intervention. It is only by extracting the enzymes and bioengineering them to degrade synthetics more rapidly, that plastivores are shown to be capable of "closing the loop of the circular economy" (Tournier et al. 2020; 2019).

It is with these biotech visions of synthetic waste efficiently excreted back into circuits of production that "good" plastivores emerge. While alone, they may seem like a promising antidote to the persistence of plastic as matter thought to be immune to biodegradation, plastic problems are not simply a stubborn solid material quality in isolation, and neither are they limited to the realm of waste management alone. Challenging such narrow framing of what plastivores "solve", I ask instead, what relations persist?

What do biotech visions of plastivore-recyclers challenge and what do they perpetuate? What is the coupling of evolution with petro-capitalism being leveraged to "naturalize"? The following section takes up these questions by turning to approaches to science and technology studies informed by feminist and queer theory, where following rather than abandoning the persistent harms of toxic chemicals and extractive capitalism becomes a means of imagining other modes of living together with plastic.

Persistent Toxicity and Microbial Kin

Even slick promises of circularity are tainted with unruly chemical complications (for an overview of the toxicity of plastics see Muncke and Zimmermann in this volume). There is speculation that the "toxic effects" of some plastic building blocks themselves limit the speed of biological degradation by inhibiting the growth of plastivorous bacteria (Espinoza et al. 2020). There are worries that microplastic producers could increase the transfer of chemical additives (Jang et al. 2018, 368), and most concerningly, evidence that their toxic effects impede the growth and oxygen-producing capacities of microbes far more broadly (Lear et al. 2021). As scientific articles become the techno-celebratory narratives of mainstream media coverage, however, chemical harm that exceeds plastivorous capacities is positioned as obligatory counterpoint, never as the crucial concern. Lingering toxicity appears as just another technical challenge to be overcome; already studies are emerging that show how yet more novel enzymes have the potential to degrade chemical additives too (Zrimec et al. 2021). For many proponents of a plastic circular economy, the goal is not dissolving synthetics back into earthy elements, but converting plastic into yet more plastic, a process which almost always requires the input of fresh plastic and chemicals along with recycled components. Even if plastics and their additives can technically be biodegraded eventually, what of the harms caused by associated extraction, production, and circulation in the meantime?

Here, approaches from feminist and queer theory are especially helpful because they counter the stubborn persistence of the status quo and attend in particular to attempts to justify relations of domination via appeals to what is natural (Chen 2012; Seymour 2018). In this case, the status quo entails the continuation of petro-capitalist production and circulation of toxins; it frames plastic problems as a matter of waste management for which tech advances taking advantage of the "natural" evolution of recycling is the fix. Bioengineering enzymes from plastivores is not a break from these relations, but arguably a means to further naturalize them in the face of a growing anti-plastics movement. Plastivores make petro-capitalism more palatable.

Plastivores, however, cannot currently biodegrade all the lingering toxic chemical additives associated with many plastics, and might never be able to do so. I take the lead from the persistent harms of chemical additives, pulling at this seemingly small fault line in biotech visions until it fractures into possibilities for something else. In doing so, I follow feminist and queer theory commitments to "embracing toxicity," a refusal to ignore violence that is at the same time a refusal to reproduce the existing social, political, and especially economic order (Chen 2012; Davis 2022). This does not mean condoning pollution or utopian tech fixes. It means following the relations of toxicity all the way through whole systems of exploitation in order to live otherwise, to "find ways of embracing the inevitable emergence of multiple strange and beautiful life forms while holding chemical companies to account for the vast harms they are enacting on numerous bodies, human and nonhuman" (Davis 2022, 83). Such an approach shifts attention to considering collective responsibilities of more-than-human relations with creativity and empathy. Regimes of domination, control, and extraction become the problem not the solution. There is no biotech fix for the violence of petro-capitalism because it is a social and political problem at the same time as a material-economic one.

Staying with the toxicity counters the default strategies of control and separation associated with current iterations of both bad and good plastic-eating. Taking plastic "off the menu" by appealing to tropes of purity that render mixing abhorrent implies an imperative to maintain synthetic-body separations by further controlling plastic materials. Similarly, the potential success of plastivore-recyclers requires the separation of enzymes from microbial communities in order to redirect plastic into a seamless loop from which materials supposedly cannot escape. In both forms, separations resonate with the violent cuts of extraction. In place of purity/containment, embracing toxicity is to dwell in the realm of contamination, where we are already intimately implicated (Shotwell 2016, 7). Bodies and environments cannot be separated; plastic is already in our blood (Alaimo 2012; Carrington 2022). As Heather Davis elaborates, "The framework of embracing toxicity—not as a good unto itself but as a refusal of the fantasy of containment, management, or barricade—allows for the potential to face futures that consist of suffering, joy, slow death, decline, survival, flourishing and hope, rather than clean breaks and ends" (Davis 2022, 98). Understanding our inextricable complicity becomes a starting point for collectively grappling with the potential of synthetic futures in all their complexities.

From this fundamental assumption of a world of interconnections, feminist and queer theory insists on broadening relationships of responsibility and care beyond the immediate and familiar. Challenging what count as "normal" relations by celebrating "improper affiliations" (Chen 2012), and expanding definitions of community beyond even species boundaries provides creative forms of resistance to the status quo by insisting on living otherwise. For Heather Davis, (2022), this resistance involves recognizing plastivorous bacteria as our "queer kin", which she argues holds the potential to upend the existing social and political order. Understood as relations of intensified responsibility that are irreducible to genetic lineage, kinship expands to new kinds of human descendants including plastic-eating microbial "offspring" (Davis 2022, 83). Davis calls on humans to assume responsibility for the co-evolution of bacteria and plastics production. Plastivores are neither simply a natural development releasing humans from obligations, nor another natural resource to exploit. Endocrine-disrupting additives that alter biological reproductive systems assume the role of interrupting the reproduction of broader social and political norms to which they are bound. In this way, "microorganisms provoke alternative conceptions of what material transformations involve [...] as an articulation of new collectives brought into the space of material politics" (Gabrys 2013, 217). This is the task of caring for emergent relations without exacerbating associated harms.

Until very recently, however, plastivores were speculative species relegated to the realms of art, science-fiction, and theory. The current materialization of plastivorous relations challenges both the temporalities of plastic degradation and the lively relationships with associated synthetic futures. For example, Yolanda Pinar's 2014 eco-art project "An Ecosystem of Excess" which includes illuminated test-tubes of life forms imagined to be emerging from plastic-saturated seas, has been read as envisioning synthetics sustaining life in a distant future without humans (Araúo 2019). Conversely, Kit Pedler and Gerry Davis's 1971 sci-fi novel *Mutant 59: the Plastic Eaters* imagined a rapid proliferation of plastic-eating bacteria threatening a world collapsing in the absence of plastic. Instead, we face the challenges of existing together with plastics and plastic-eating, and the concrete challenges that must be addressed in realizing other ways of living. At the same time that the material endurance of plastic is questioned, we are faced with the urgency of contending with futures that emerge seemingly already appropriated. How can we hold on to the optimism exemplified by Jennifer Gabrys approaching plastivorous bacteria as metabolic "carbon workers" whose "material residues provide fodder for rethinking the trajectories of our material economies and ecologies, outside the closed-loop of renewed capital," (Gabrys 2013, 224), when their labor is already enlisted in the service of capitalist production rather than subverting it? Similarly, how can we consider our responsibilities to microbial kin when they are presented as already reduced to isolated enzymes? The emergence of plastivores now points to the specific challenges—but also necessity of—insisting on relationships as a form of resistance to the violent separations that constitute extractive logic.

Conclusions

Plastic's persistence is not the result of a material quality alone. It cannot be reduced to the toughness of synthetic chemical bonds that some enzymes may be coerced into breaking. The problems and promises of plastic-eating cannot be evaluated in narrow petro-capitalist terms of economic (non)productivity. Rather, plastic's persistence is a reminder of whole sets of relations that endure, from chemical residues to economically convenient material myths of seamless plastics recycling (see also Eckert in this volume), from appeals to nature to technoscience in the service of petro-capitalism. That microbes developing the capacity to metabolize plastic is given as evidence of ecological resilience and adaptation is particularly cause for alarm because it is loaded with the potential to displace responsibility onto natural processes. As Rob Nixon cautions in his account of slow violence, "the deep-time thinking that celebrates natural healing is strategically disastrous if it provides political cover for reckless corporate short-termism" (Nixon 2011, 22). With the unexpected speed of plastivores' evolution understood as opening plastic up to and even compressing the temporalities of "natural" healing, plastivores are enlisted to smooth over the increasingly problematic toxic accumulations of synthetic plastic production. Plastic-eating solves a problem for petro-capitalism, not the problem of it.

Tracing lineages of persistence through the emergence of plastivore science can teach us about how to live—and live better together—in the plastics age if they are considered beyond a narrow technoscience framework. Plastivore relations can help counter the continued over-emphasis on waste as the privileged point of intervention into plastic problems. From Keep America Beautiful's "litterbug" campaign, to industry-sponsored clean- ups and recycling competitions, focusing on individual responsibility for consumer waste has long-served corporate interests (Lerner 2019). What is helpful about visions of a plastic circular economy, is the potential to position waste-production relationships as a crucial site of intervention. Doing so requires that we think beyond technical critiques of biotechnical processes, to consider how the funneling of waste back into production is just the latest play in the contest for the ability to keep polluting. That unprecedented recent commitments to ratifying a legally binding global treaty for reducing plastic pollution takes aim at the entire plastic lifecycle suggests meaningful shift from waste management at the level of policy. That "reference to concern over chemicals in plastic was taken out of the agreement after objections from delegations including the United States" (Tabuchi 2022), suggests chemical harm does indeed hold the potential to unravel dreams of circularity.

Resisting the current trajectories of synthetic futures cannot happen only in theory. Imagining and realizing alternative relations for life in the Plastics Age cannot be the work of humanities scholars alone. Researchers in the natural sciences can also grapple with their relationships to petro-capitalism, just as queer and feminist theorists are attending to chemical harm. Following the lead of Max Liboiron's (2021) anti-colonial science laboratory, how might we normalize reconfiguring science and humanities research processes: tracing where funding comes from and the systems it perpetuates; considering the pollution research may produce and when to abstain; and questioning whether knowledge is inherently "good" in itself or only in its relations. How might collective science-humanities articulations of "broader impacts" help attend to systemic persistences? Let's rise to the challenge of forming radical collaborations capable of subverting them, together.

References

Alaimo, Stacy (2012). States of suspension: Trans-corporeality at sea. *ISLE: Interdisciplinary Studies in Literature and Environment*, 19 (3), 476–493.
Ap, Tiffany (2016). New plastic-eating bacteria could help save the planet. CNN March 11. 10.09.2021 https://www.cnn.com/2016/03/11/world/bacteria-discovery-plastic/index.html.
Bundy, Marie H, and Henry A. Vanderploeg (2002). Detection and capture of inert particles by calanoid copepods: the role of the feeding current. *Journal of Plankton Research*, 24 (3), 215–223.
Campanale, Claudia, Carmine Massarelli, Ilaria Savino, Vito Locaputo, and Vito Felice Uricchio (2020). A detailed review study on potential effects of microplastics and additives of concern on human health. *International Journal of Environmental Research and Public Health*, 17 (4), 1212.
Carpenter, Scott (2021). The race to develop plastic-eating bacteria. Forbes. March 10th. 10.09.2021 https://www.forbes.com/sites/scottcarpenter/2021/03/10/the-race-to-develop-plastic-eating-bacteria/?sh=771efee77406.
Carrington, Damian (2022). Microplastics found in human blood for first time. The Guardian, March 24th. 26.05.2022 https://www.theguardian.com/environment/2022/mar/24/microplastics-found-in-human-blood-for-first-time.
Chen, Mel (2012). *Animacies: Biopolitics, racial mattering, and queer affect*. Durham: Duke University Press.
Cornwall, Warren (2021). Could plastic-eating microbes take a bite out of the recycling problem? Science, July 1st. 10.09.2021 https://www.science.org/content/article/could-plastic-eating-microbes-take-bite-out-recycling-problem.
Davis, Heather (2022). *Plastic matter*. Durham: Duke University Press.
de Araújo, Vanbasten (2019). Life without humankind'—queer death/life, plastic pollution, and extinction in an ecosystem of excess. *Women, Gender & Research*, 3 (4), 49–61.
De Wolff, Kim (2017). Plastic naturecultures: Multispecies ethnography and the dangers of separating living from non-living bodies. *Body & Society*, 23 (3), 23–47.
Eckert, Sandra (2022). Circularity in the plastic age: Policymaking and industry action in the European Union. In Johanna Kramm and Carolin Völker (eds.). *Living in the plastic age. Perspectives from humanities, social sciences and environmental sciences*. Frankfurt, New York: Campus.
Gabrys, Jennifer (2013). Plastic and the work of the biodegradable. In Jennifer Gabrys, Gay Hawkins, and Mike Michael (eds.). *Accumulation: the material politics of plastic*. New York: Routledge.
Gabrys, Jennifer, Gay Hawkins, and Mike Michael (2013). *Accumulation: the material politics of plastic*. New York: Routledge.

Haram, Linsey E, James T. Carlton, Gregory M. Ruiz, and Nikolai A. Maximenko (2020). A plasticene lexicon. *Marine Pollution Bulletin*, 150, 110714.

Hawkins, Gay (2006). *Ethics of waste*. Oxford: Rowman & Littlefield.

Hohn, Donovan (2011). *Moby Duck: The true story of 28,800 bath toys lost at sea and of the beachcombers, oceanographers, environmentalists, and fools, including the author, who went in search of them*. New York: Viking Press.

Irigoien, Xabier (2018). Introduction to virtual special issue: zooplankton plastivory. *Journal of Plankton Research*. https://academic-oup-com.libproxy.library.unt.edu/plankt/pages/zooplankton_plastivory.

Jang, Mi. Won Joon Shim, Gi Myung Han, Young Kyoung Song, and Sang Hee Hong (2018). Formation of microplastics by polychaetes (*Marphysa sanguinea*) inhabiting expanded polystyrene marine debris. *Marine Pollution Bulletin*, 131, 365–369.

Lear, G, J. M. Kingsbury, S. Franchini, V. Gambarini, S. D. M. Maday, J. A. Wallbank, L. Weaver, and O. Pantos (2021). Plastics and the microbiome: impacts and solutions. *Environmental Microbiome*, 16 (1), 1–19.

Lerner, Sharon (2019). Waste only: How the plastics industry is fighting to keep polluting the world. The Intercept. July 20th. 10.09.2021 https://theintercept.com/2019/07/20/plastics-industry-plastic-recycling/

Liboiron, Max (2021). *Pollution is Colonialism*. Durham: Duke.

Mulhern, Owen (2021). "Plastic eating bacteria: a viable solution to the plastic problem?" Earth.Org March 15th. 10.09.2021 https://earth.org/data_visualization/plastic-eating-bacteria-a-viable-solution-to-the-plastic-problem/.

Muncke, Jane, and Lisa Zimmermann (2022). Chemicals in plastics: Risk assessment, human health consequences, and political regulation. In Johanna Kramm and Carolin Völker (eds.). *Living in the plastic age. Perspectives from humanities, social sciences and environmental sciences*. Frankfurt, New York: Campus.

Nixon, Rob (2011). *Slow violence and the environmentalism of the poor*. Cambridge: Harvard University Press.

Pedler, Kit, and Gerry Davis (1971). *Mutant 59: the plastic eaters*. London: Souvenir Press.

Seymour, Nicole (2018). *Bad environmentalism*. Minneapolis: Minnesota University Press.

Shotwell, Alexis (2016). *Against purity: Living ethically in compromised times*. Minneapolis: Minnesota University Press.

Simon, Matt (2021). Baby poop is loaded with microplastics. Wired, September 22nd. 03.03.2022 https://www.wired.com/story/baby-poop-is-loaded-with-microplastics/.

Smith-Llera, Danielle (2020). *You are eating plastic every day*. North Mankato Minnesota: Compass Point Books.

Tabuchi, Hiroko (2022). The world is awash in plastic. Nations Plan a treaty to fix that. New York Times. March 2nd. https://www.nytimes.com/2022/03/02/climate/global-plastics-recycling-treaty.html.

Yoshida, Shosuke, Kazumi Hiraga, Toshihiko Takehana, Ikuo Taniguchi, Hironao Yamaji, Yasuhito Maeda, Kiyotsuna Toyohara, Kenji Miyamoto, Yoshiharu Kimura, and Kohei Oda (2016). A bacterium that degrades and assimilates poly(ethylene terephthalate). *Science*, 351 (6278), 1196–1199.

Zettler, Erik R, Tracy J Mincer, and Linda A. Amaral-Zettler (2013). Life in the plastisphere: Microbial communities on plastic marine debris. *Environmental Science & Technology*, 47 (13), 7137–7146.

Zrimec, Jan, Mariia Kokina, Sara Jonasson, Francisco Zorrilla, and Aleksej Zelezniaka (2021). Plastic-degrading potential across the global microbiome correlates with recent pollution trends. *MBio*, 12 (25), e02155-21.

Contributors

Gisela Böhm is professor in psychology at the University of Bergen, and adjunct professor at Inland Norway University of Applied Sciences, Lillehammer. Her research interests are risk perception and decision making in the context of environmental risks, with a focus on risk, morality, and emotional responses to such risks.

Nicole Brennholt works in the water sector at the North Rhine-Westphalian State Agency for Nature, Environment and Consumer Protection. Previously, she worked as a research scientist at the German Federal Institute of Hydrology, where she was in charge of (inter)national projects dealing with (micro-)plastic issues. She holds a doctorate in ecology.

Ana Isabel Catarino (PhD) studies the impact of anthropogenic activities on aquatic organisms, with a special focus on plastic pollution effects and global change. She is a researcher at the Flanders Marine Institute, Belgium, and a co-PI of the project COLLECT, Citizen Observation of Local Litter in Coastal ECosysTems.

Kim De Wolff is an Assistant Professor of Philosophy at the University of North Texas. She is co-editor of Hydrohumanities: Water Discourse and Environmental Futures, and brings a feminist science and technology studies approach to entanglements of plastics, oceans and petro-capitalism.

Rouven Doran is associate professor in general psychology at the University of Bergen. His research interests are centered on individual and social factors that shape perceptions and responses to environmental issues, including topics such as sustainable consumption and energy transition.

Sandra Eckert holds a Chair in Comparative Politics at the Friedrich-Alexander-Universität Erlangen-Nürnberg. She previously was Marie Skłodowska-Curie COFUND fellow and Associate Professor at the Aarhus Institute of Advanced Studies and Assistant Professor at Goethe University

Frankfurt. Her research covers the fields of comparative public policy and European Integration.

Gert Everaert (PhD) is the leader of the research group "Ocean and Human Health" at the Flanders Marine Institute, Belgium. His research focuses on the understanding of how the ocean impacts human well being and how anthropogenic activities potentially alter the marine environment, including biodiversity and ecosystems functioning.

Paula Florides has a background in political science. In cooperation with the Institute for Social-Ecological Research in Frankfurt, she has conducted research on European plastics policy for her bachelor's thesis. Based in Cyprus, she currently focuses on environmental peacebuilding and intersectional environmentalism with a special interest in art as activism.

Maja Grünzner holds a MSc in social psychology and is an Early Stage Researcher within the H2020 LimnoPlast ITN project researching behavioral approaches to the microplastics problem and its potential solutions. She is a PhD candidate and part of the Environmental Psychology Research Group at the University of Vienna.

Maria Pia Herrling is the head of the Nanoparticle Group and innovation project manager at Ovivo Switzerland AG with the main focus on nanoparticle detection and control in ultrapure water for semiconductor industry and product development. She holds a diploma degree in environmental science and received her PhD in chemical engineering, both at Karlsruhe Institute of Technology, Germany.

Maren Heß is a biologist with focus on ecotoxicology at North Rhine-Westphalia Office of Nature, Environment and Consumer Protection. She works on monitoring and evaluating chemical pollutants in surface waters, including microplastics as emerging contaminants.

Henner Hollert is Head of the Department Evolutionary Ecology & Environmental Toxicology (E3T) at Institute of Ecology, Diversity and Evolution, Goethe University Frankfurt. He is head of the working group on bioassays in the European Norman Network and one of the co-authors of the report on effect-based tools of the European Commission.

Alexander Hooyberg is a doctorate student at the Flanders Marine Institute and Ghent University, Belgium. He studies the impact of exposure to coastal environments and litter on psychological health. His multidisciplinary research topic combines biology with knowledge from health sciences, psychology, and sociology.

Thorsten Hüffer is an analytical and environmental chemist at the Department of Environmental Geosciences (University of Vienna). His research focuses on the fate and transport of particulate and dissolved contaminants in environmental systems. He heads the expert committee "Plastics in the aquatic environment" within the German Water Chemistry Society.

Natalia P. Ivleva is leading Raman & SEM Group at Technical University of Munich (TUM), Institute of Hydrochemistry (IWC) Chair of Analytical Chemistry at the TUM. Her research interests focus on the analysis of complex environmental and industrial samples by means of Raman microspectroscopy, surface-enhanced Raman scattering, and stable isotope approach, with special attention on online/high-throughput analytics.

Jutta Kerpen is a professor for waste water and water treatment at the RheinMain University of Applied Sciences in Wiesbaden. Her research focuses on the analysis of microplastics in industrial and municipal waste water and the development and improvement of analytical procedures for microplastics in different environmental matrices.

Johanna Kramm holds a doctorate in human geography and is a junior research group leader at the ISOE-Institute for Social-Ecological Research in Frankfurt am Main. She currently works on plastic waste at the interface of science and policy. Her research is informed by approaches of political ecology and new materialisms.

Christian Laforsch, Chair of Animal Ecology I at the University of Bayreuth, has twelve years of experience in microplastics research. Current expertise ranges from the identification of microplastics from various environmental matrices with a focus on method development and toxicological impacts of microplastics from the molecular to the ecosystem level.

Martin Löder is head of the plastics group at the Chair of Animal Ecology I at the University of Bayreuth. His research focuses on the topic of microplastics, especially on method development for sampling, purification and identification of microplastics in environmental samples and on the ecological implications of microplastics.

Jane Muncke holds a doctorate in environmental toxicology and is the managing director and chief scientific officer of the Food Packaging Forum in Zürich, Switzerland, working at the science-policy interface on chemicals in food packaging. She is a full scientific member of the Society of Toxicology (SOT), the American Chemical Society (ACS), the Society of Environmental Toxicology and Chemistry (SETAC), and the Endocrine Society.

Sabine Pahl is a Professor of Urban and Environmental Psychology at the University of Vienna. She is doing basic and applied research, which mainly focuses on the human dimension in environmental issues. She has provided science advice into policy at national, European and international levels, contributing psychological and behavioral science perspectives.

Luca Raschewski is a sociologist and currently works on their PhD-thesis on the role of ignorance regarding microplastics risks in everyday practices at the University of Aachen.

Marcos Felipe-Rodriguez holds a master's degree in cognitive psychology and neuropsychology and is a doctoral researcher at the University of Bergen under a Horizon2020 grant. He currently researches on environmental psychology, investigating mental models and risk perception and communication in regards to plastic pollution.

Marine Severin is a doctorate student at the Flanders Marine Institute in Ostend, Belgium, with affiliation at Ghent University and KU Leuven. She has an academic background in clinical psychology and is now researching underlying mechanisms in the relationship between coastal environments and mental well-being.

Sabrina Schiwy is team leader of the Department of Evolutionary Ecology & Environmental Toxicology (E3T) at the Institute of Ecology, Diversity and Evolution at Goethe University Frankfurt, Germany. She is an expert in aquatic ecotoxicology and has experience in the assessment of micropollutants, micro-and nanoplastics in sediment and water.

Markus Schmitz is a PhD candidate at the department Evolutionary Ecology & Environmental Toxicology at Goethe University in Frankfurt am Main, Germany. He is experienced in in-vivo bioanalytical methods in fish and currently investigates the ecotoxicological effects of tire abrasion and road stormwater runoff in a large-scale bioassay battery.

Immanuel Stieß is head of the research unit "Energy and Climate Protection in Everyday Life" at the ISOE-Institute for Social-Ecological Research in

Frankfurt. He is a sociologist and planning expert and received his doctorate in architecture, urban planning and landscape planning. His focus is on social-ecological life-style research in nutrition as well as construction and housing and indicators of sustainability for social life.

Georg Sunderer works as a freelance researcher at the ISOE-Institute for Social-Ecological Research in Frankfurt. He studied sociology at the universities of Trier, Germany and Galway, Ireland. His major fields of study are environmental attitudes and behavior, ethical consumption, and the acceptance of social-ecological innovations.

Carolin Völker holds a doctorate in ecotoxicology and is junior research group leader at the ISOE-Institute for Social-Ecological Research in Frankfurt. She works on environmental risk assessment and systemic risks of microplastics.

Stephan Wagner is head of the research group "Anthropogenic Water Cycles" at Hof University of Applied Sciences, Germany. He holds a PhD in water technology. Aiming at the development of safe and sustainable materials his research focuses on environmental analysis and the fate and environmental impact of chemicals and particles

Lisa Zimmermann holds a doctorate in ecotoxicology and is a scientific communication officer at the Food Packaging Forum in Zürich, Switzerland. Her current research and communication efforts focus on harmful chemicals in food packaging, bioplastics, and human health impacts of microplastics.

Index

accumulate 65, 71, 208, 224
accumulation 7, 15, 34, 52, 68, 70, 72, 133, 158, 264
additive 70
agenda-setting 13, 25, 26, 30, 41–43, 241, 246, 256
analytical techniques 55
aquatic environment 14, 58, 63, 82, 86, 129, 193, 279
awareness 8, 16–18, 23, 24, 33, 34, 89, 91–93, 100, 101, 111, 112, 115–117, 121, 126, 133–135, 137–139, 150–152, 162–165, 167, 169, 171–173, 178, 181, 185, 212, 224
awareness-behavior gap 18, 171
bacteria 75, 214, 262, 266, 268, 270, 273, 274
ban 20, 31, 35, 122, 123, 125, 240, 242–244, 250, 252
behavior 13, 17, 18, 23, 74, 86, 91, 104, 110, 113, 117, 120, 127, 151, 157, 159, 162, 169, 170–175, 178, 180, 183–195, 223, 265, 281
behavior change 13, 18, 120, 151, 169, 170, 172, 173, 178, 180, 185, 186, 188, 189, 192, 194, 195
bioassay 52, 73, 203, 280
bioavailability 73

biodegradable plastics 51, 126, 129, 216, 217, 227, 229, 244, 253, 254
biodegradation 67, 216, 217, 226, 267
bioplastic 197
biotechnological fix 264
bisphenol A
BPA 70, 202, 225, 226
business 20, 133, 166, 223, 240, 241, 254
business models 223, 240
business power 20, 241
capitalism 13, 21, 263, 264, 267–269, 271, 272, 277
chemical composition 24, 54, 61, 68, 197, 203, 204, 223, 229, 233
chemical mixture 208, 221
chemical recycling 239, 249, 257
chemical regulation 243, 248
circular economy 13, 14, 18, 20, 23, 40, 41, 157, 197, 212, 216, 218, 222, 226, 227, 237, 239, 246, 247, 252, 254–260, 262, 264, 267, 268, 272
circularity 9, 20, 25, 236, 244, 247, 249, 251, 253, 258, 263, 264, 268, 272
citizen science 14, 17, 23, 93, 108, 134–139, 141, 148–157, 159, 160–167

climate change 24, 39, 91, 94–96, 103–110, 170, 190, 194, 195, 247–249, 256
complexity 10, 14, 15, 20, 52, 53, 58, 68, 72–77, 87, 199, 214, 222, 223
Comprehensive Action Determination Model 173
consumer 11, 16, 24, 35, 42, 43, 75, 91, 104, 110, 111, 115, 119, 122, 128, 129, 166, 169, 189, 198, 203, 220, 229, 233, 243, 250, 263, 272
consumption 7–9, 18, 21, 41, 67, 91, 101, 111, 113, 118, 120, 121, 169, 170, 176–178, 181, 183, 187–189, 191, 193, 204, 218, 223, 224, 227, 244, 246, 253, 266, 277, 281
co-optation 236
Davis 21, 23, 49, 85, 160, 166, 264, 269, 270, 273, 274
definition of plastics 52
degradation 32, 53, 61, 67, 80, 114, 121, 197, 200, 216, 217, 219, 225, 228, 231, 263, 268, 270
demographic
  socio-demographic 18, 91, 102, 104, 152, 155, 158
density 56, 58, 65, 68, 74, 149
detection 55, 57, 61, 62, 74, 79, 82, 185, 278
distribution 17, 34, 52, 56, 57, 63, 65, 74, 80, 94, 116, 134, 157, 163, 166, 216, 224, 267
diversity 10, 15, 17, 23, 53, 58, 59, 68, 156, 164
dog poop bags 16, 114, 116, 125–127
down-cycling 20, 239
eating 21, 181, 262, 263, 265, 266, 269–271, 273, 274

ecological 10, 15, 22, 55, 61, 72, 75, 76, 103, 106, 132, 159, 173, 174, 190, 191, 217, 246, 257, 271, 279, 281
ecology 55, 277, 279
ecosystem 9, 71, 75, 76, 133, 273, 279
ecotoxicological 55, 61, 72, 75–78, 280
ecotoxicology 10, 11, 55, 72, 77, 278, 280, 281
emission 15, 58, 62, 63, 215
endocrine 83, 202, 203, 207, 214, 223, 228, 231, 232, 242
endocrine-disrupting 207, 223
end-of-life 214, 243, 252
endpoint 15, 72
environmental awareness 93, 117
environmental groups 235, 241, 242, 249
environmental policy 236, 246, 257
enzymes 21, 263, 267–269, 271
EU plastics policy 43
European Commission 13, 25, 31, 32, 34–42, 44–46, 48, 72, 80, 89, 90, 93, 105, 148, 160, 171, 190, 220, 236, 239, 243, 244, 246–251, 253–256, 259, 278
European Strategy for Plastics in a Circular Economy 25, 41, 46
everyday life 10, 23, 111, 112, 121, 128, 130, 132
exposure 19, 67, 68, 71, 81, 85, 90, 94, 95, 109, 154–156, 159, 197, 202, 204, 205, 207–210, 212, 214, 222, 223, 226, 231, 232, 235, 239, 243, 248, 266, 278
extraction
  extract 30, 53, 56, 149, 235, 263, 264, 268, 269
fleece jackets 16, 115, 124, 125, 126

Food Contact Materials
  FCMs 197, 211, 227
food packaging 131, 197–199,
  201, 203–208, 210–212, 219,
  220, 223, 225–227, 229, 231,
  252, 280, 281
fossil-based 20, 113, 235, 240, 253
fragmentation 51, 54, 62, 63, 67,
  78, 85, 216
freshwater 51, 63, 71, 74, 75, 79,
  82, 84–86, 100, 109, 160, 170
FTIR 55, 57, 59, 82
Great Pacific Garbage Patch 34,
  38
green taxonomy 247, 249
habitat 75, 76, 78, 83
habits 18, 111, 112, 123, 125, 171,
  174, 176, 185
hazard 77, 88–90, 94–96, 99,
  101, 169, 201, 203–205, 212,
  229, 231
hazardous 41, 109, 133, 201, 204,
  205, 214, 217, 218, 220, 221,
  224, 232, 242, 248, 254
health risks 9, 80, 93, 205, 212,
  239
hormones 207
human health 9, 10, 17, 79, 81,
  93, 96, 98, 101, 107, 135, 153,
  177, 190, 201–203, 206, 207,
  210, 211, 214, 217, 223, 230,
  232, 242, 244, 258, 273, 274,
  281
identification 58, 60, 80–82, 84,
  184, 229, 230, 279
in vitro 24, 73, 74, 202, 203, 214,
  230, 233
in vivo 73, 74, 202
incineration 246, 250
indirect effects 70
ingestion 31, 35, 42, 68, 70, 80,
  263, 265
interventions 183
invasive species 75

kin 21, 270, 271
Kingdon 13, 22, 26, 37, 41, 43, 48
knowledge 9, 16, 23, 32, 34, 48,
  51, 52, 74, 82, 90–92, 95,
  101–104, 108, 112, 113, 115,
  117, 126, 128, 135, 139,
  150–153, 157, 158, 161, 170,
  207, 222, 223, 254, 258, 259,
  261, 262, 272, 279
leachate 74
legitimacy 241
Liboiron 161, 263, 265, 272, 274
life cycle 39, 189, 194, 215, 224,
  226, 235, 236, 242, 249
linear economy 9, 20, 237, 238,
  253
macroplastic 133
marine litter 13, 20, 30–32, 34,
  36–38, 41–43, 108, 134, 136,
  139, 152, 153, 161–163, 166
mass media 35, 42, 111
material recovery 240
mechanical recycling 249, 250,
  257
Meikle 7, 8, 23
mental models 16, 88, 99,
  100–102, 280
microplastics 8, 10, 11, 13–16,
  20–24, 32, 35, 42, 43, 51–58,
  60–65, 67–73, 75–86, 88, 100,
  103–110, 112–129, 131, 148,
  159, 161, 163, 170, 177, 180,
  181, 186, 188, 191, 193, 194,
  216, 218, 246, 247, 266, 273,
  274, 278–281
migration 198, 200, 201, 205, 206,
  208, 222, 225, 229, 232
mitigation 32, 87, 95, 98, 99, 105,
  134, 190, 247, 249, 256
mixture toxicity 15, 72, 203, 210
moist toilet paper 16, 115, 122,
  123, 126, 127
monomer 74, 202, 208, 215

Multiple Streams Framework 13, 26, 29, 47, 48, 50
nanoplastics 51, 52, 54, 59, 71, 73, 77, 82–84, 86, 280
ocean literacy 17, 135, 150, 151, 157, 159, 162, 167
oil use 253
oligomers 70, 206, 219
organism 35, 42, 71, 72, 203, 232, 266
packaging waste 220, 224, 240, 246, 251, 256
participation 17, 18, 93, 134, 135, 138, 140, 148–152, 155, 156, 159, 160, 161, 166, 184
particles 8, 14, 15, 24, 30, 32, 33, 51–57, 59–62, 66–68, 70–80, 84, 86, 88, 100, 101, 113, 265, 273, 281
peelings 16, 115, 116, 121, 122, 126, 127
persistence 15, 53, 72, 76, 224, 264, 267, 268, 271, 272
petrochemical industry 21, 253
phthalates 8, 31, 35, 70, 202, 229, 242, 243, 248
physical effects
physical impacts 70, 77
Plastic Age 3, 4, 7, 8, 11, 12, 23, 44, 235, 262
plastic industry 247
plastic pollution 9, 10, 16, 18–22, 24, 34, 41, 63, 64, 83, 89, 90, 95, 106–09, 131, 133–137, 139, 148–150, 151, 154–157, 159–162, 165, 167, 169–172, 183, 185, 188–190, 192, 193, 216, 218, 224, 258, 262, 265–267, 272, 273, 277, 280
plastic soup 35, 38, 42
plasticene 253, 259, 274
plastivore 263–265, 268, 269, 272
policy process 26, 236, 259
policy-industry nexus 20, 236, 247

political science 244, 252, 278
pollutant 25, 32
pollution 14, 18, 21–23, 25, 31, 33–35, 49, 63, 64, 80, 90, 95, 100, 101, 103, 104, 106, 107, 111, 133–135, 148, 150, 151, 153, 154, 157, 158, 160, 161, 166, 169–171, 177, 178, 183, 188, 190, 191, 193, 224, 232, 247, 254, 256, 265, 267, 269, 272
polymer 14, 15, 52–54, 56, 58, 61–63, 67, 72, 74, 80, 85, 203, 204, 216, 218, 242
polymerization 8, 19, 70, 200, 206, 217
post-consumer waste 243
Potočnik 32, 36–42, 45, 48, 49
practices 7, 10, 16, 17, 23, 75, 94, 107, 111–113, 115, 120, 121, 126–128, 130, 158, 160, 164, 193, 262, 263, 280
problem framing 13, 27, 34, 42
product groups 31, 113, 116
public mood 35, 36, 42
PVC 20, 31, 35, 80, 203, 213, 221, 236, 242, 243, 247, 248, 250, 252–254, 259, 260
Pyr-GC 56, 61
quantification 58, 61, 80, 81, 148, 231
queer theory 21, 268, 269, 270
Raman spectroscopy 55, 57, 59, 60
REACH 78, 221, 243, 248
recyclable 181, 187, 244, 246, 251
recycling 14, 20, 21, 25, 31, 40, 41, 44, 81, 131, 134, 175, 183, 186, 189, 193, 197, 216, 217, 219–222, 224, 231, 235, 238, 241, 242, 244, 246–251, 253, 256–259, 262–264, 266, 267, 269, 271–275

reduce 14, 18, 20, 38, 77, 78, 80, 84, 89, 91, 100, 105, 113, 123, 125, 127, 128, 130, 161, 177, 178, 181, 188, 189, 193, 204, 208, 215, 218, 240, 244, 246, 250
regulation 20, 31, 75, 78, 79, 178, 184, 192, 197, 211, 220, 221, 225, 227, 236, 237, 241, 242, 247–250, 254, 257–259, 274
reproduction 70, 202, 270
resource efficiency 14, 39, 40, 42, 251, 258
responsibility 14, 17, 18, 21, 31, 37, 39, 41, 42, 98, 100, 101, 104, 112, 118, 119, 122, 125–127, 161, 172, 173, 189, 190, 224, 246, 263, 270–272
restrict 244, 246, 249
reuse 25, 41, 176, 218, 219, 221, 222, 240, 246, 251
risk 8, 10, 13, 14, 16, 19, 22–24, 42, 51, 77–79, 86–99, 101–104, 106–110, 114, 155, 169, 170, 191, 197, 201, 205–212, 221, 223, 228, 229, 230, 236, 242, 277, 280, 281
risk assessment 14, 51, 77, 78, 114, 197, 201, 205–208, 210, 211, 242, 281
risk perception 8, 10, 13, 16, 23, 42, 88, 92, 94, 96, 102–104, 106, 107, 110, 169, 191, 277, 280
safety 90, 107, 201, 204, 211, 212, 214, 215, 217–223, 225, 227, 229, 260
sampling 15, 30, 53, 56, 57, 65, 66, 115, 134, 138, 140, 148, 149, 151, 157, 160, 279
Science and Technology Studies 12
secondary material 239, 248

sediment 53, 57, 65, 67, 71, 73, 74, 84–86, 149, 163, 280
sewage sludge 64
sewage system 64, 122
Shove 111, 113, 132
single-use plastic 19, 36, 38, 43, 171, 172, 176, 246, 252
size 9, 15, 53–59, 62, 65, 67, 68, 71, 72, 74, 75, 115, 200, 223, 224
social practice 132
socio-economic 17, 92, 155, 156, 158, 263
solution 10, 11, 18, 21, 28, 40–42, 106, 121, 123, 131, 161, 169, 188, 189, 216, 218, 222, 224, 262, 264, 267, 269, 274
source 62, 80, 83, 89, 100, 101, 117, 125, 131, 165, 197, 208, 215, 253, 262, 266
stakeholder involvement 20, 241
surface 14, 53, 56–58, 64–68, 71, 72, 75, 80, 115, 160, 161, 200, 278, 279
sustainability 9, 20, 193, 214, 215, 217, 227, 236, 239, 242, 248, 249, 253, 256, 259, 281
technology 11, 21, 104, 119, 120, 195, 223, 239, 250, 263, 268, 277, 281
TED-GC 56, 61
toxic 21, 70, 77, 78, 83, 86, 203, 221, 235, 242, 255, 256, 268, 269, 272
toxicity 10, 24, 37, 70, 73, 74, 77, 78, 83, 86, 90, 202–204, 206, 207, 214–216, 222, 224, 225, 228, 229, 233, 263, 268, 269
transport 8, 51, 57, 63, 68, 82, 111, 135, 139, 253, 279
uncertainty 10, 16, 77, 87, 96, 110, 121, 206, 225, 228
use phase 242
vector 71, 75, 82, 83

virgin material 239
voluntary agreements 240, 248, 250
waste 7−9, 13, 14, 19, 21−23, 25, 31, 37−42, 44, 51, 63, 83, 94, 100, 101, 108, 111, 114, 123, 126, 127, 131, 133, 139, 149, 150, 162, 166, 172, 183, 185, 187, 192, 212, 214, 216, 218−221, 224, 226, 227, 231, 236, 239, 240, 242, 243, 244, 246−248, 250−252, 254−259, 263, 267, 269, 272, 274, 279
waste policy 31, 38, 39, 40
wastewater 16, 61, 63−65, 86, 104, 114, 122, 125, 127
water column 15, 53, 57, 65, 72
water phase 66
water surface 57, 65
weathering 67, 68, 166, 229
window of opportunity 28, 30, 31, 37, 39, 41, 42